AI for Real Estate:
How Artificial Intelligence will Change Real Estate

AI for Real Estate
How Artificial Intelligence will Change Real Estate
Michael Fielden

Published by:
Support Geeks, Inc. San Jose, California

Copyright ©2024 Michael Fielden
All rights reserved.

All rights reserved. No part of this publication may be reproduced, stored in retrieval system, or transmitted in any form or by any means, electronic, mechanical, photocopying, recording, scanning, or otherwise, except as permitted under section 107 or 108 of the 1976 United States Copyright Act without the prior written permission of the author.

Limit of Liability and Disclaimer of Warranty: While the publisher and author have used their best efforts in preparing this book, they make no representation or warranties with repect to the accuracy or completeness of the contents of this book and specifically disclaim any implied warranties or merchantability or fitness for a particular purpose. The advice and strategies contained herein may not be suitable for your situation. You should consult with a professional where appropriate.

Paperback ISBN: 979-8-9897704-0-3
Audiobook ISBN: 979-8-9897704-1-0
Kindle ASIN: B0CQMMQ56Q

Acknowledgements

I would like to thank everyone that read and reread this book to help me make sure that it turned out just right. An extra special acknowledgement and thanks to Geoffrey Schierbaum, Vern McGeorge, John Bennison and Matthew Antonelli.

Of course, I'd also like to give thanks to my family, Gina, Sofi, Natalie, and Tucker the dog, for their tolerance and love.

I would like to thank OpenAI for providing me with access to its GPT-4 language model and other resources, which were invaluable in the writing of this book.

I would like to thank Midjourney for providing me with the AI image generation tool that enabled me to create the cover image for this book. Their innovative technology and powerful AI algorithms played a significant role in bringing the visual concept for my book's cover to life.

If you're reading this on printed paper, you and I both can be thankful to Keith at Lightning Press in Santa Clara, California.

Contents

Introduction
The Rise of AI in Various Industries
The Potential of AI in the Real Estate Sector

Understanding AI: A Brief Overview — 24
History and Development of AI
AI vs Traditional Computing

The Need for AI in Real Estate — 37
The Current Challenges in the Real Estate Industry
How AI Can Address These Challenges

AI in Property Search and Listings — 42
AI-driven Chatbots for 24/7 Customer Inquiries

Price Predictions and Market Analysis — 50
AI for Comparative Market Analysis (CMA)
Anticipating Market Shifts Using AI

The Rise of Virtual Property Tours — 59
Enhancing Property Visualization with AR
Benefits for Real estate agents and Clients

AI for Paperwork and Legal Procedures — 69
Predictive Analytics for Assessing Property Legal Risks
Automating Repetitive Paperwork Tasks

Client Relationship Management and AI — 79
Personalizing Client Interactions Using AI Insights
Predicting Client Needs and Preferences

Improving Marketing and Outreach — 87
Chatbots for Lead Generation
Social Media Insights and Optimization Using AI

Challenges and Ethical Considerations — 95
Bias in AI Algorithms and its Implications
The Human Touch vs. AI: Striking a Balance

Getting Started with AI — 105
List of Popular AI Tools for Real estate agents
Training and Support for Implementing AI Solutions

The Future of Real Estate with AI — 113
Preparing Your Business for an AI-Driven Future

Embracing the Future with AI — 120

Appendix A — 122

Appendix B — 125

Appendix C — 128

Appendix D — 130

Appendix E — 132

Appendix F — 134

Appendix G — 136

Introduction

In the vast landscape of industries undergoing radical transformation due to technology, real estate stands as one of the most intriguing sectors on the brink of a revolution. Historically, the real estate market has been an arena grounded in personal relationships, handshake deals, and traditional methods. But as with all sectors, evolution is inevitable.

"AI for Real Estate" seeks to bridge the gap between the age-old principles of real estate and the cutting-edge advancements in Artificial Intelligence. At first glance, the worlds of AI and real estate might seem miles apart. Yet, on closer inspection, the synergy becomes unmistakably clear. From predictive analytics forecasting market trends to virtual tours powered by augmented reality, AI is not just knocking on the door of the real estate industry; it's ready to redesign the entire house.

> But what does this mean for today's real estate professional? Is AI a tool to be leveraged, or is it a looming threat to traditional practices? How can real estate agents adapt to and thrive in this new landscape, ensuring they remain not just relevant but indispensable?

This book will embark on a journey through the intricate web of AI, dissecting its components and deciphering its implications for the real estate industry. Whether you're a seasoned real estate veteran, a newcomer to the field, or someone with a keen interest in the convergence of technology and property, this guide aims to illuminate the path forward.

The future of real estate is unfolding, and it's intertwined with the marvels of AI. Let's embark on this journey together, unlocking opportunities, dispelling fears, and charting a course for a future where real estate agents and AI don't just coexist but collaborate in harmony.

The Rise of AI in Various Industries

The story of Artificial Intelligence isn't confined to one industry. It's a narrative that unfolds across myriad sectors, reshaping them in profound and often unprecedented ways. Before delving into AI's role in real estate, it's essential to understand its broader impact. This understanding may be easier to see in the way that AI has impacted other industries.

Healthcare: From Diagnosis to Treatment

The realm of medical diagnostics has undergone a transformative shift with the advent of AI. Machine learning algorithms, equipped with the power to process vast amounts of data, have proven adept at interpreting intricate diagnostic imagery. What's more impressive is that, in certain instances, these algorithms demonstrate a level of accuracy that surpasses the discerning capabilities of the human eye.

Apart from refining diagnostic techniques, AI has introduced the healthcare industry to the wonders of predictive analytics. This technology mines patient data to foresee potential health risks, paving the way for proactive medical strategies. By catching potential health concerns in their nascent stages, medical professionals can institute timely interventions, often preventing the escalation of minor issues into major complications.

Beyond the diagnostic domain, AI's influence permeates the treatment landscape. With its data-driven analysis, AI crafts treatment plans that are meticulously tailored to individual patient profiles. This ensures that patients aren't just receiving generic care but are beneficiaries of medical attention specifically designed for their unique physiological and genetic makeup.

Retail: Personalizing Shopping Experiences

Have you ever found yourself astonished by how e-commerce platforms appear to precisely pinpoint what you've been yearning to purchase? This uncanny ability isn't mere coincidence; it's the sophisticated craftsmanship of AI-driven recommendation engines. These advanced systems delve deep into the digital footprints you leave behind, scrutinizing every click, scroll, and hover.

Drawing insights from your browsing habits, previous purchases, and even subtle preferences, these engines curate a selection of products that align with your tastes and needs. This not only amplifies the shopping experience for users by presenting them with apt choices but also holds immense value for e-commerce platforms. By ensuring that users are presented with products they're likely to be interested in, these platforms increase the likelihood of sales, striking a harmonious balance between consumer satisfaction and business profitability.

Banking and Finance: Smarter Decisions, Safer Transactions

AI has revolutionized the banking sector, offering improvements and efficiencies across various facets of the industry. One of the most visible advancements is the use of chatbots for customer service. These AI-powered chatbots are capable of handling a wide range of customer queries, from simple account inquiries to more complex transaction-related questions, offering quick and efficient responses that improve customer experience and reduce the workload on human staff.

In the realm of security, AI algorithms play a critical role in detecting and preventing fraudulent transactions. By analyzing patterns in large volumes of transaction data, these algorithms can identify anomalies that might indicate fraudulent activities, such as unauthorized access or unusual spending patterns. This not only helps in safeguarding customers' assets but also enhances trust in banking institutions.

Moreover, the advent of robo-advisors has been a game-changer in personal finance management. These AI-driven tools assist individuals in making informed investment decisions by analyzing vast datasets that include market trends, historical performance, and individual financial goals. Robo-advisors can offer personalized investment advice and portfolio management services, traditionally accessible only to high-net-worth individuals, to a broader range of customers. This democratization of financial advice helps individuals optimize their investments and better plan their financial futures.

AI in banking also extends to credit scoring, where machine learning models assess creditworthiness more accurately and inclusively than traditional methods. By considering a wider range of factors, including non-traditional data points, AI can offer a more nuanced understanding of a borrower's risk profile.

AI has not only enhanced customer-facing services in banking but has also bolstered security measures, democratized financial advice, refined credit assessments, and streamlined internal operations; thereby reshaping the banking landscape into a more efficient, secure, and customer-centric industry.

Transportation: The Road to Autonomy

Self-driving cars, a concept that was once considered futuristic, are rapidly becoming a tangible reality, largely due to advancements in artificial intelligence (AI). At the core of this technological breakthrough are sophisticated algorithms capable of processing vast amounts of data from a multitude of sensors embedded in the vehicle. These sensors include cameras, radar,

LiDAR (Light Detection and Ranging), and ultrasonic sensors, all of which continuously gather critical data about the vehicle's surroundings.

The AI algorithms integrate and analyze this data in real-time, enabling the vehicle to make informed and precise decisions. This process involves recognizing and interpreting traffic signals, identifying obstacles like pedestrians, other vehicles, and roadblocks, and understanding complex traffic patterns and behaviors. The AI system also needs to predict the actions of other road users and react accordingly, which requires an advanced level of situational awareness and predictive analysis.

Entertainment: Curated Content and Creation

Streaming giants such as Netflix and Spotify have integrated advanced AI mechanisms to elevate the user experience. Whenever you log in and find that new show, movie, or song suggestion that resonates with your taste, it's the intricate AI algorithms diligently working behind the scenes. By examining your viewing or listening history, as well as your interaction patterns with the platform, these systems are able to curate recommendations that align seamlessly with your preferences.

Yet, the influence of AI on the world of streaming doesn't stop at recommendations. Innovators are pushing the boundaries by harnessing AI for content creation itself. This includes everything from generating music compositions that captivate the listener to assisting in scriptwriting processes for films and shows. As AI continues to evolve, it holds the promise of reshaping not just how we consume content, but also how it is conceived and brought to life.

Agriculture: Farming in the Digital Age

Precision agriculture employs artificial intelligence (AI) to enhance crop yields and optimize the use of resources. This innovative approach integrates advanced technologies to create a more efficient and productive agricultural process. Drones, equipped with AI-driven cameras, play a pivotal role in this system. They are deployed over fields to monitor various conditions, providing a comprehensive view of the crops and their environment. This high-resolution imagery is crucial for assessing the health and growth of the plants.

The data collected by the drones is then analyzed using sophisticated machine learning algorithms. These algorithms process the information to extract meaningful insights about the field conditions. Based on this analysis, AI systems can offer precise recommendations for planting, watering, and harvesting. This targeted approach ensures that each area of the field receives the exact care it needs, significantly improving the efficiency of farming operations. The result is an agricultural process that is not only more productive but also more sustainable, reducing waste and conserving resources.

Ever noticed how things are getting smarter and faster around us? That's thanks to AI, a kind of super-brain that's helping many businesses do better and dream bigger. Just like a Swiss Army knife, AI can be adjusted to fit different jobs, turning our wildest imaginations into reality. Let's dive in and see how this magic touch of AI can spruce up the world of real estate!

The Potential of AI in the Real Estate Sector

Imagine a world where buying a house or renting an apartment is as easy and smooth as ordering your favorite food online. That's the exciting future we're looking at when we combine the powers of AI and real estate! With AI in the mix, property deals can happen faster, choices become a lot smarter, and the entire journey feels like a breeze. And here's the thrilling part: as technology keeps getting better and better, the world of real estate isn't just trying to catch up—it's aiming to be a front-runner in all these cool advancements. For all the real estate agents out there who are ready to hop on this exciting ride, the road ahead isn't just filled with sunshine; it's set to rewrite the way we think about property altogether!

Here are some areas in real estate where AI can truly shine:

Predictive Analytics: Forecasting Market Trends

Remember the times when real estate agents would sift through old records and mostly trust their intuition to make decisions? Those days seem like a distant memory now, thanks to the wonders of AI! Today's real estate world has taken a leap into the future with AI-driven predictive analytics. Instead of just looking in the rearview mirror at past data, AI helps agents peek into the future. It can predict upcoming twists and turns in the market, like when house prices might go up or down or when a particular neighborhood is about to become the next hot spot. This isn't just fancy tech talk—it means real estate agents can now plan their moves way ahead, ensuring they're always a step ahead in the game. With such powerful tools at their fingertips, agents can serve their clients better, making smarter and more informed decisions than ever before.

Virtual Tours: Viewing Homes from the Living Room

Virtual reality (VR) backed by the intelligence of AI is rapidly changing the face of agents showing property. Gone are the days when homebuyers would spend weekends hopping from one property to another. With the advent of VR, they can now embark on immersive journeys through potential homes from the comfort of their living rooms.

These aren't just simple walkthroughs either. AI-powered VR

tours offer a depth of detail that feels incredibly lifelike. From the texture of the hardwood floors to the play of sunlight through the windows, everything is rendered with meticulous precision. And it doesn't stop there. Imagine being able to change the color of the walls or swap out furniture with a simple click. Homebuyers can now personalize spaces in real time, testing different decor styles and color palettes to see what aligns best with their vision. This blend of technology not only saves time and energy but also offers a richer, more personalized property-hunting experience, making dream homes come alive even before the purchase.

Chatbots and Virtual Assistants: 24/7 Customer Service

Timely responses are crucial to ensuring customer satisfaction. Gone are the days when customers would patiently wait for days or even hours for a reply to their queries. They expect, and rightfully so, immediate and accurate answers. This is where AI-driven chatbots come into the picture, becoming an indispensable tool for businesses, especially in the real estate sector.

These chatbots are not just simple automated responders; they're equipped with intelligent algorithms that enable them to understand and address a wide range of common questions posed by potential buyers and sellers. Whether it's inquiries about property specifications, neighborhood details, or the process of setting up a property visit, these bots can provide the necessary information swiftly.

Moreover, their 24/7 availability ensures that customers from different time zones or those who prefer late-night browsing still receive the assistance they need. By reducing wait times and providing accurate, round-the-clock support, AI chatbots significantly enhance the overall customer experience, building trust and loyalty in the process.

AI-Driven Property Management

The integration of AI into property management is revolutionizing the way landlords and property managers oversee and maintain their properties. Instead of reactive measures, which often come after a problem has already occurred, AI enables proactive steps by predicting potential maintenance issues. For instance, by analyzing patterns and data from appliances and systems, AI can forecast when a particular piece of equipment might fail or need servicing. This not only reduces the chances of unexpected breakdowns but also extends the life of the equipment, saving on replacement costs in the long run.

Additionally, one of the significant challenges in property management has always been energy consumption. AI helps in optimizing energy usage by learning patterns of consumption and adjusting systems accordingly. For instance, smart thermostats, like those available from Nest, Honeywell, Ecobee and others, can learn the preferences of the occupants and adjust heating or cooling for optimal comfort while minimizing wastage. Lighting, too, can be optimized based on occupancy, ensuring spaces are not unnecessarily lit.

For landlords and property managers, these advancements mean streamlined operations, reduced overhead costs, and fewer emergency call-outs. On the flip side, tenants reap the benefits of these innovations in the form of a more comfortable, responsive, and efficient living environment. Imagine living in a space where the ambient temperature is always just right, where potential issues are addressed even before they manifest, and where energy bills are kept at a minimum due to intelligent optimizations. This is the promise AI brings to property management, making life better for both landlords and tenants.

Enhanced Decision Making

The real estate industry has always been data-intensive, but the dawn of AI has opened up avenues to process and analyze this data like never before. Traditionally, matching properties to potential buyers was a manual process, relying heavily on a real estate agent's intuition and experience. While these elements remain crucial, the introduction of AI adds an extra layer of sophistication and accuracy to the process.

By analyzing vast datasets, AI can understand subtle patterns and correlations that might be missed by the human eye. For instance, beyond just considering the number of bedrooms or proximity to schools, AI algorithms can take into account a myriad of factors, such as local noise levels, proximity to parks, historical appreciation rates, and even finer details like the orientation of the house concerning sunlight. When combined with specific buyer preferences, such as their hobbies, family size, or even lifestyle habits, the system can suggest properties that are a near-perfect match for the buyer's unique needs and desires.

For real estate agents, this level of precision is a game-changer. It means fewer property viewings that end in disappointment, and more that conclude with a successful sale. This not only saves time for both the real estate agent and the buyer, but also reduces the associated costs of multiple viewings. Moreover, a quick and efficient matching process results in faster sales, which is beneficial for both the seller and the agent.

From the client's perspective, the experience becomes significantly more enjoyable and less tedious. Instead of spending weekends touring properties that aren't quite right, they can focus on a curated list that closely aligns with what they're looking for. This personalized approach leads to a higher level of client satisfaction, fostering trust and often resulting in repeat business and referrals.

In essence, the fusion of AI with traditional real estate practices is creating a win-win scenario. Real estate agents benefit from more streamlined operations and higher success rates, while buyers and sellers enjoy a more tailored and efficient experience.

The confluence of AI and real estate promises a future where transactions are swifter, choices are smarter, and experiences are seamless. As technology continues its march forward, the real estate sector is poised not just to adapt but to lead in many innovations. For real estate agents ready to embrace this change, the future isn't just bright; it's revolutionary.

Understanding AI: A Brief Overview

What is AI? Artificial Intelligence, commonly referred to as AI, is a multifaceted and rapidly evolving field in computer science that focuses on creating machines capable of thinking and acting like humans. But let's unpack this definition a bit more.

At its core, AI aims to replicate or simulate human intelligence in machines. This doesn't just mean mimicking human behaviors but also replicating our unique ability to learn, reason, perceive our surroundings, solve problems, and even potentially understand emotions.

There are several key components to AI:

Learning: Just as humans learn from experiences, AI is designed to learn from data. Machine Learning, a subset of AI, allows machines to improve from exposure to data over time, refining their algorithms and decision-making capabilities.

Reasoning: AI systems are designed to solve problems through logical deduction. They can make decisions that help them achieve specific goals, much like a chess player deciding on the next move.

Self-correction: A crucial aspect of human intelligence is our ability to recognize when we've made a mistake and to learn from it. Similarly, a well-designed AI system can acknowledge its errors and refine its algorithms to improve future decisions.

Problem-solving: Beyond the realm of traditional reasoning and decision-making, artificial intelligence systems have the capability to be meticulously trained to navigate through their operational environments, effectively solving complex problems that span a wide array of domains. These AI systems can process vast amounts of data and execute tasks with a level of speed and accuracy that often surpasses human capabilities, demonstrating their efficiency in both rapid problem-solving and delivering highly accurate results, often outperforming humans in tasks that require extensive data analysis or pattern recognition.

AI isn't just about software. When people envision AI, they often picture robots, and indeed, robotics is a significant field within AI. Robots use AI algorithms to perform tasks, interact with their environment, and even engage with humans in increasingly sophisticated ways. Robots is not what this book is about, but if they take off in real estate, that will be next.

A common misconception is equating AI with basic automation. While automation can be a product of AI, not all automated systems possess the learning and reasoning capabilities inherent to true artificial intelligence.

In the context of real estate and many other industries, AI's potential lies in data analysis, predictive modeling, customer interaction, and even virtual property showcases. By understanding AI's essence, real estate agents can better harness its power, paving the way for innovative solutions in property marketing, sales, and management.

History and Development of AI

Long before computers were even a concept, humans have been fascinated with the idea of artificial beings endowed with intelligence. Ancient civilizations, from the Greeks with tales of automatons like Talos, to Chinese legends of mechanical men, have captured imaginations with stories of inanimate objects brought to life.

Yet, the true intellectual foundations for AI were laid during the Age of Enlightenment. Philosophers began dissecting the nature of human thought and cognition, pondering if these could be replicated mechanically. This contemplation laid the conceptual foundations that would, centuries later, influence the development of AI.

Fast forward to the mid-20th century, when British mathematician Alan Turing posed the provocative question:

> "Can machines think?" Turing's 1950 paper introduced the Turing Test, suggesting that a machine could be considered 'intelligent' when its actions were indistinguishable from a human's.

It was in 1956, during a seminal workshop at Dartmouth College, that the term "Artificial Intelligence" was coined, and the field of AI research was born. Spearheaded by visionaries like John McCarthy and Marvin Minsky, this period was marked by soaring optimism and the development of the first AI programs.

The subsequent decades witnessed fluctuating fortunes for AI:

1960s: This period saw significant funding and the development of foundational AI programs, such as the General Problem Solver and ELIZA, a primitive chatbot. Expectations were high, with many researchers predicting that machine intelligence would soon eclipse human intelligence.

1970s: The challenges of translating theoretical AI into practical solutions became more evident. Funding began to dry up, leading to the first "AI winter," a period of stagnation and reduced interest in AI research.

1980s: The development of expert systems, which emulated human decision-making in specific domains, rekindled interest. Yet, by the end of the decade, the limitations of these systems and massive costs led to another AI winter.

21st Century: Renaissance and Rapid Advancements

With the dawn of the new millennium, the AI spring began. Two main catalysts were the exponential growth in computational power and the vast amounts of data available. Machine learning, and more notably, deep learning with neural networks, became central to AI research.

Key milestones of this period include:

IBM's Deep Blue defeated the world chess champion in 1997, marking a significant milestone in the field of artificial intelligence. Following this, the development of voice-activated assistants like Siri and Alexa revolutionized the way humans interact with technology, making it more intuitive and personalized. These developments collectively demonstrate the rapid advancement and growing capabilities of AI in various complex tasks.

Core Principles of AI

AI is broad, and its definitions vary based on context. However, two core principles run central:

> **Learning**: Much like humans learn from experiences, AI systems learn from data. This learning can be supervised (they're trained on a specific dataset) or unsupervised (they identify patterns and relationships within data).
>
> **Reasoning**: AI systems can solve problems through logic. They can take a problem, break it down, analyze possible solutions, and choose the optimal one.

Subfields of AI

AI is an umbrella term, encompassing a plethora of subfields:

Machine Learning (ML): ML algorithms improve through experience. They're trained on data, from which they derive insights and make decisions.

Neural Networks: Inspired by human biology and our neural system, these are algorithms designed to recognize patterns. They interpret sensory data and cluster raw input, labeling and categorizing it.

Deep Learning: A subset of ML, deep learning uses multi-layered neural networks. It's behind many contemporary AI marvels, from voice assistants to facial recognition.

Natural Language Processing (NLP): Enabling machines to understand and respond in human language, NLP powers chatbots, translation services, and more.

Robotics: This field deals with machines capable of a series of actions, making physical interactions with the world.

Expert Systems: These are computers that mimic human experts. They make decisions based on a set of rules and knowledge.

Vision or Computer Vision: Machines interpret and decide based on visual data, just like human vision.

The Emergence OpenAI and Simple Public Access to AI

OpenAI was founded in December 2015, emerging from a unique vision shared by its founders, including Elon Musk, Sam Altman, Greg Brockman, Ilya Sutskever, Wojciech Zaremba, and John

Schulman, among others. This group of entrepreneurs and AI researchers came together with a shared concern about the potential risks of artificial intelligence and a belief in the importance of developing AI in a way that would benefit all of humanity.

The founding of OpenAI was driven by the recognition that AI technology has the potential to bring about significant changes in society. The founders believed that in order to harness the full benefits of AI while mitigating its risks, it was essential to conduct research in an open and collaborative environment. This philosophy stood in contrast to the more closed and proprietary research environments typical of many tech companies and academic institutions.

From its inception, OpenAI positioned itself as a research organization dedicated to achieving the highest ideals of AI development. Its mission was not just to advance the field of artificial intelligence, but to do so in a manner that prioritizes safety, transparency, and broad accessibility. The founders committed to freely sharing much of their research with the public, aiming to democratize access to AI technologies and ensure their benefits were widely distributed.

To support its ambitious goals, OpenAI initially operated as a non-profit organization. This structure allowed it to focus on long-term research objectives without the pressure of generating immediate financial returns. It attracted some of the brightest minds in AI research, who were drawn to the organization's mission and approach to AI development.

Over time, OpenAI evolved in response to the changing landscape of AI research and the need for sustainable funding to support its extensive research projects. In 2019, it transitioned to a "capped-profit" model with the creation of OpenAI LP, a limited partnership subsidiary under the overarching non-profit OpenAI Inc. This new structure was designed to attract external investment while maintaining a strong commitment to OpenAI's original ethos of openness and public benefit. The capped-profit model allows OpenAI to generate revenue through commercializing its technologies, but caps the returns to investors, ensuring that the primary focus remains on advancing AI in a way that is safe and beneficial for everyone.

Throughout its evolution, OpenAI has made significant contributions to the field of AI, including the development of groundbreaking technologies like GPT (Generative Pretrained Transformer) series of language models and DALL-E, an AI system capable of generating highly realistic images from textual descriptions. These innovations, along with its commitment to ethical AI development, have positioned OpenAI as a leading organization in the field, shaping the trajectory of AI research and its applications in society.

AI vs Traditional Computing

Comparing Artificial Intelligence (AI) with Traditional Computing offers a revealing lens through which we can witness the remarkable progression of technology over the years. Traditional Computing, which is rule-based and relies on explicit programming, has served as the foundation for many technological innovations. It involves step-by-step processing, with an emphasis on structured algorithms and deterministic outputs. On the other hand, AI, particularly its modern subset, machine learning, functions by analyzing vast amounts of data to discern patterns, make decisions, and even predict future outcomes. Instead of being explicitly programmed for a specific task, AI systems learn from the data they are fed, making them more adaptable and capable of handling complex, unstructured information.

The implications of these advancements have been particularly transformative for sectors like real estate. Traditional Computing enabled the first wave of digitalization in the industry, streamlining processes such as property listing, database management, and financial transactions. However, with the advent of AI, the real estate industry has been introduced to a host of new possibilities. AI-driven tools can now predict property value fluctuations, offer personalized property recommendations based on user preferences, and even automate tasks like virtual property tours using augmented reality.

By dissecting the capabilities and constraints of both AI and Traditional Computing, we can derive crucial insights. For instance, while Traditional Computing offers reliability and a set framework, AI brings flexibility and a level of personalization. But AI also comes with challenges like the need for large datasets and the potential for biases in predictions. These insights can greatly influence how technology is employed in real estate, ensuring that the best of both worlds are leveraged.

Understanding the unique strengths and potential pitfalls of each technological approach is vital. It not only highlights the vast potential they bring to real estate but also sheds light on areas of improvement. Such awareness paves the way for future innovations, ensuring that the real estate industry remains at the forefront of technological evolution, capitalizing on the synergies of both AI and Traditional Computing.

Understanding Traditional Computing

Before diving into the various forms of computing, it's important to have a clear grasp of traditional computing. This type of computing is straightforward and rule-based. It operates by following specific, predefined instructions, which are known as algorithms. These algorithms are like a set of step-by-step directions that the computer follows to accomplish tasks. Whether it's performing basic arithmetic calculations or running complex software applications, traditional computing relies on these unchanging instructions to function.

In the context of real estate, traditional computing has played a transformative role. It has successfully transitioned many tasks from being manual and paper-based to being digital and automated. This shift has brought about significant efficiency and accuracy improvements. For instance, in real estate, traditional computing systems are used to manage large databases. These databases can store vast amounts of information about properties, clients, transactions, and more, all in an organized and easily accessible way.

Moreover, traditional computing aids in financial aspects of real estate. Complex calculations needed for mortgage amortization, property valuation, and investment return analyses, which once required manual computation or the use of simple calculators, can now be performed swiftly and accurately using computer software. This reduces the chance of errors and saves considerable time.

Another area where traditional computing has made a substantial impact is in document creation and management. Real estate involves a plethora of documents - contracts, lease agreements, sale deeds, and more. Creating these documents manually is not only time-consuming but also prone to errors. With traditional computing, real estate professionals can use templates and word processing software to create, edit, and manage these documents more efficiently. It also allows for the digital storage of these documents, making retrieval and sharing easier and more environmentally friendly than paper-based systems.

Overall, traditional computing has provided a solid foundation for numerous functions within the real estate sector, enhancing productivity, reducing errors, and improving overall management.

Artificial Intelligence: A Paradigm Shift

Conversely, AI goes beyond the fixed rules used in traditional computing. It introduces systems that can learn on their own, adapt to new situations, and even make decisions. A key part of AI is machine learning. This allows a computer system to learn from data by itself. Instead of being given specific instructions for every task, it observes patterns and details in the data it receives and improves or changes how it works based on what it learns.

In the world of real estate, AI shows up in many ways and does a lot more than just organize data. For example, it's used in predictive analytics, which involves analyzing data to predict future trends and outcomes. This can help real estate professionals understand things like which areas might become more popular or valuable. AI is also used in chatbots and virtual assistants. These tools can talk to customers, answer their questions, and help them in real-time, making customer service faster and more efficient.

AI also helps in making decisions. For instance, it can analyze lots of market data to help real estate agents and investors make informed decisions about buying, selling, or pricing properties. This use of AI not only makes data handling easier but also improves how customers are interacted with and helps in making better, more informed decisions.

Comparative Exploration: AI vs Traditional Computing

Learning and Adaptability:

Traditional Computing: Adheres to specified algorithms, lacking the capability to learn or adapt.

AI: Exhibits learning and adaptability, modifying its behavior based on data and experiences.

Decision-making:

Traditional Computing: Incapable of making decisions independently, relying on defined protocols.

AI: Capable of decision-making by leveraging learned patterns and information.

User Interaction:

Traditional Computing: Limited to user-directed interactions, operating strictly based on user input.

AI: Enables more natural interactions, understanding and even anticipating user needs through learned patterns.

Complex Problem Solving:

Traditional Computing: Solves problems based on clear, logical steps provided by algorithms.

AI: Can address complex, ambiguous problems, finding solutions even when problem-solving pathways are not clearly defined.

Data Handling and Analysis:

Traditional Computing: Manages and processes data based on pre-defined functions.

AI: Not only manages data but also extracts insights, identifies patterns, and even predicts future trends.

Ingraining AI into Real Estate: Beyond Traditional Norms

When we compare Artificial Intelligence (AI) to Traditional Computing, we see how technology has evolved. Traditional Computing follows set rules and specific programming. It's a step-by-step process that has powered many tech advancements. In contrast, AI, especially machine learning, works by analyzing lots of data to find patterns and make predictions. Rather than being programmed for one task, AI learns from the data it receives, making it versatile and able to handle varied information.

This tech evolution has had a big impact on sectors like real estate. Traditional Computing started the digital shift in this field, making things like property listings and transactions easier. But with AI's entry, there are even more possibilities. AI tools can predict property prices, suggest properties tailored to a user's taste, and even give virtual tours using augmented reality.

When we look closely at both AI and Traditional Computing, we learn important things. Traditional Computing is reliable and follows a set pattern, while AI is more flexible and personalized. However, AI requires large amounts of data and might sometimes be biased. Knowing this can shape how we use technology in real estate, combining the best parts of both.

It's essential to recognize the strengths and weaknesses of these technologies. This not only shows their value in real estate but also areas they can improve. Being aware of this helps innovate for the future, ensuring real estate keeps benefiting from both AI and Traditional Computing.

Challenges and Ethics

As AI systems became more integrated into daily life, concerns arose. Issues of bias in AI models, ethical considerations in AI decision-making, and the broader implications of AI on job markets, privacy, and society have taken center stage.

AI's journey is filled with challenges, both technical and ethical:

> **Bias and Fairness**: AI models reflect the data they're trained on. If this data contains biases, the AI's decisions could also be biased.

Transparency: Understanding how AI models make decisions, especially complex neural networks, is challenging. This "black box" nature raises concerns about accountability.

Job Impacts: As AI systems become more proficient, there are fears about job displacements across sectors.

Security: AI systems can be vulnerable to attacks. Adversarial attacks, where the AI gets deceptive input leading it to misclassify information, are a rising concern.

Conclusion: Coalescing Two Worlds

Combining AI and the usual computer systems in the real estate business can create smart, straightforward, and clever ways of working. It's really important for real estate agents to understand how both of these tech tools work, along with what they're good and not so good at. This helps agents make the most of both technologies as everything moves more and more online.

In simpler words, by using the trusty and exact methods of regular computing and mixing them with the forward-thinking and flexible abilities of AI, real estate agents can not only improve how they work day-to-day but also find new, creative ways to give their clients a better and more modern service in a world that's becoming more digital by the minute.

The Need for AI in Real Estate

This chapter will explain how important it is to use AI technology in real estate. The idea is not to replace the work done by humans, but to help improve it. With AI, the real estate industry can more easily overcome the challenges it faces today.

The Current Challenges in the Real Estate Industry

The real estate industry is constantly changing, always adapting to new challenges that arise. Currently, the industry faces several obstacles that impact both real estate professionals and their clients. This includes managing a huge amount of property data, effectively responding to client questions, accurately valuing properties in a changing market, and making sure that everything from paperwork to transactions is done smoothly and without mistakes.

The Current Landscape of Challenges in Real Estate

> **Digesting a World of Information**: Picture this, a library with thousands of books, with every piece of information about every property you're dealing with - this is the kind of data real estate agents sift through daily. Managing this mammoth amount of data - from client details, and property specifics to the undulating trends in the market, has become an uphill battle. Imagine having a tool that instantly sorts, analyzes, and provides exactly the data you need, when you need it. This is one of the spaces where AI can truly shine, turning a mountain of data into easily digestible, actionable insights.
>
> **Meeting Sky-High Customer Expectations**: In an era where instant gratification is often the norm, clients expect quick, accurate, and personalized responses. Maintaining a high standard of customer service, which entails being perpetually available and providing precise information promptly, can be daunting. AI steps in here with capabilities like chatbots, that

can interact with clients any time of the day, providing instant responses and information, thereby elevating the standard of customer service offered.

Navigating Through Property Valuations: When it comes to determining the value of properties, a myriad of factors come into play, and with market conditions that are constantly in flux, achieving accurate valuations can be a perplexing task. Here, AI isn't just a tool; it's a companion that can sift through numerous data points, ensuring that the valuations provided are not just accurate but are also reflective of real-time market conditions.

Ensuring Operations are Smooth Sailing: Handling documentation, managing schedules, communicating with clients, and ensuring transactions are error-free - the operational side of real estate is undeniably multifaceted. AI can automate many of these processes, ensuring that tasks are completed efficiently, timely, and without error, allowing real estate professionals to dedicate more time to client interaction and strategic planning.

Understanding that we are not replacing real estate professionals with machines but providing them with a tool that amplifies their capabilities is vital. It's about melding the analytical, data-driven capabilities of AI with the emotional intelligence, relationship-building, and strategic planning capabilities of humans. This synergy doesn't just alleviate the challenges currently faced by the industry but also paves the path toward a future where decisions are data-informed, operations are smooth, and client experiences are enriched.

By recognizing the challenges and embracing AI as a robust tool to navigate through them, the real estate industry can not only continue to flourish but also provide an enriched, efficient, and evolved experience to both professionals and clients alike.

How AI Can Address These Challenges

AI as the Beacon of Resolution

In today's fast-paced world, the tasks of managing properties, engaging with clients, and handling extensive paperwork present significant challenges for real estate professionals. These responsibilities require a high degree of skill and efficiency, akin to performing a complex balancing act. This context sets the stage for exploring how Artificial Intelligence (AI) – sophisticated technologies capable of human-like thinking and learning – can provide substantial support. AI has the potential to streamline these complex tasks, offering stability and efficiency in the multifaceted role of a real estate agent, thereby simplifying their professional life and enhancing overall productivity.

Clarifying How AI Can Help

AI as Your Personal Assistant: Think of AI as having a super-efficient, never-tiring assistant who can manage loads of tasks without breaking a sweat. From scheduling your appointments, sending out reminders to clients, to even answering their basic queries at any time of the day (or night!), AI can ensure that your operations run seamlessly while you focus on building relationships with your clients.

Cutting Through the Data Jungle: Navigating through the vast jungle of data, from property details, client information, to ever-changing market trends, can be overwhelming. AI acts as a proficient guide, helping you sift through this information jungle by sorting, analyzing, and presenting precisely what you need. Imagine being able to predict which properties will be in demand or what will be the next big trend in the real estate market! With AI, these insights are not just dreams but accessible realities.

Chatbots: Your 24/7 Customer Service: We live in an age where waiting is passé, and instant replies are the new norm. AI-powered chatbots can be your round-the-clock customer service, addressing queries, providing information, and ensuring that your clients always feel heard and valued, even when you're catching up on some well-deserved sleep.

AI's Eyes on Market Trends: Picture this: you're able to accurately predict the ups and downs of the market or know exactly when a property's value will skyrocket. AI enables this by continuously analyzing market trends and helping you stay one step ahead, ensuring that your strategies are always in line with the pulse of the market.

Smart Homes and AI's Magic Touch: AI isn't just simplifying backend operations but is also enhancing property appeal through smart homes. Imagine walking into a house that adjusts its lighting, temperature, and even music to suit your mood! For a potential buyer, a smart home isn't just a property; it's an experience, and AI makes creating this charm effortless and efficient.

Keeping Documents in Check: With properties, come piles of documents. AI can manage and organize these documents, ensuring that you're always on top of your paperwork, and never miss out on details, be it client preferences, transaction histories, or legal documents.

The Ethical Compass and AI

In today's data-driven world, possessing vast amounts of data comes with a significant amount of responsibility. Artificial Intelligence (AI) plays a crucial role in ensuring that this responsibility is upheld. AI systems are designed to manage and process information ethically and responsibly, maintaining the utmost confidentiality. This involves implementing robust security measures to ensure that your client's data remains secure and well-protected against any potential breaches. By doing so, AI not only safeguards the data but also helps in building and strengthening a relationship with clients that is deeply rooted in trust and reliability. This trust is essential, as it reassures clients that their sensitive information is handled with the highest level of care and professionalism.

Navigating through the real estate industry with its varying challenges might seem like a daunting task. But with AI by your side, managing data, ensuring excellent customer service, keeping an eye on market trends, enchanting clients with smart homes, and ensuring ethical data management become less of a challenge and more of a smooth sail.

In the pages that follow, we'll dive deeper into each of these aspects, exploring how you, as a real estate agent, can integrate AI into your operations, ensuring that the services you provide are not just efficient and up-to-date but are also reflective of a future-forward mindset in the realm of real estate.

As we delve deeper into the world where AI and real estate converge, remember: AI isn't here to replace the agent; it's here to enhance, enrich, and simplify, ensuring that the challenges faced today are smoothly transformed into the steppingstones for a technologically enhanced future of real estate.

AI in Property Search and Listings

Using AI for Property Recommendations: Imagine having a friend who listens to your every need, understands your wishes without you explicitly stating them, and then, voila, conjures a list of properties that match your dreams perfectly! That friend, in the expansive world of real estate, is no one else but Artificial Intelligence (AI). Let's explore this intriguing journey where AI takes the varied aspirations of diverse clients and transforms them into a list of homes that aren't just structures but dreams crafted into reality.

> **Understanding the Unsaid with AI:** Real estate agents are skilled at understanding client needs, but what if we could amplify that understanding manifold? AI dives deep into previous client interactions, searches, and preferences, even picking up on those unsaid desires, to present a selection that's as close to their dream property as possible. It's not merely about the number of rooms or locality; it's about understanding their lifestyle, their future plans, and their unseen aspirations.
>
> **AI as Your Property Matchmaker:** Marrying client preferences with the available properties can sometimes be a complex puzzle. Here's where AI comes into play, working tirelessly, scanning through numerous listings, and aligning them with client preferences, ensuring every match it makes is one step closer to their dream home. It's like having a matchmaker who not only understands what you desire but also knows where to find it!
>
> **AI Can Search in New Ways:** AI systems excel in processing structured data — information that is organized in a predictable pattern, like prices, locations, property sizes, and number of rooms. By harnessing this data, AI can create detailed profiles of properties, comparing and contrasting them with the specific preferences and requirements of a buyer. For instance, a family looking for a spacious, four-bedroom house within a

specific budget can be instantly matched with suitable listings through an AI-driven platform. This level of customization is achieved through sophisticated algorithms that not only understand the explicit criteria set by the buyer but also learn from their interactions and preferences, refining search results over time.

Moreover, AI's capability to analyze pictures transforms the home-buying experience. Through advanced image recognition and processing techniques, AI can extract valuable information from property photos, providing insights that go beyond the standard specifications. It can analyze room layouts, identify features like swimming pools or garden spaces, and even assess the condition of the property. This visual analysis, combined with structured data, offers a comprehensive view of each listing. For example, if a buyer is interested in bright, airy spaces, AI can prioritize homes with large windows and open layouts in its suggestions.

Dynamic Pricing with AI's Insight: AI doesn't just stop at finding the right property but goes a step further by ensuring that the pricing is apt, considering the current market trends, property features, and locality benefits. It assures that you and your clients are always in the winning spot, striking deals that are fair and intelligent.

Immersive Experiences Through Virtual Tours: With AI, property viewing transcends geographical boundaries. Virtual tours powered by AI offer immersive experiences to clients, allowing them to explore properties from the comfort of their homes. It's not just about viewing the property; it's about experiencing it, understanding the space, the ambiance, and visualizing their life in it.

Effortless Document Management: Once a property catches your client's eye, the subsequent steps involve numerous documents. AI ensures that every paper, be it related to property details, client interaction, or transaction history, is organized and easily accessible. So, each step including the property selection, until the keys are handed over, is smooth and hassle-free.

Predictive Analysis: A Peep into the Future: AI also enables agents to provide a futuristic view to their clients. Through predictive analysis, it can provide insights into the property's future value, helping clients make informed decisions that are not just beneficial today but continue to be so in the future as well.

Creating Stories with AI: Every property has a story and AI helps you tell it in a way that resonates with the client. By understanding their preferences, AI can highlight aspects of the property that align with their desires, creating a narrative that's not just about a house but about creating a home.

In this chapter, we have shown a peek into how AI, with its myriad of capabilities, acts as a bridge between real estate agents and clients, ensuring that every property recommendation represents what clients are looking for and every transaction is accurate and complete.

As we step into subsequent chapters, we will explore further how AI, with its analytical and empathetic capabilities, continues to revolutionize the real estate landscape, ensuring that every transaction is not just a commercial exchange but an emotional and intelligent connection between the buyer, seller, agent, and property.

Predictive Analytics to Identify Hot Properties

Navigating the real estate market effectively demands strategic decision-making and impeccable timing. The challenge lies in identifying properties that have the potential to become highly desirable before they attract widespread attention. This is where the integration of Artificial Intelligence (AI) and Predictive Analytics becomes invaluable. These technologies serve as advanced tools for real estate professionals, offering a significant competitive edge.

AI and Predictive Analytics use vast amounts of data, including market trends, demographic shifts, economic indicators, and historical property values, to forecast future market movements. By analyzing this data, these tools can pinpoint properties that are likely to appreciate in value or become in-demand, based on factors like location, price trends, and community developments. This foresight enables real estate agents to advise their clients more effectively, helping them to invest in properties that are poised for growth or to sell properties at the most opportune time.

Furthermore, AI algorithms can personalize property recommendations for clients, matching them with listings that align with their preferences and investment goals. This not only streamlines the property search process but also enhances customer satisfaction by providing tailored advice.

In essence, AI and Predictive Analytics transform the real estate landscape by empowering agents with data-driven insights. This leads to more informed decision-making, better client service, and the potential for higher returns on property investments. As the real estate market continues to evolve, these technologies will likely become even more integral to the success of real estate professionals.

Clarification on How AI Predictive Analytics May Work

Painting the Future Picture with Data: Predictive analytics refers to utilizing data, statistical algorithms, and AI to identify the likelihood of future outcomes based on historical data. In the realm of real estate, it helps agents predict which properties are set to become notably desirable or gain value in the forthcoming days or months. It's like having a superpower where you can forecast the future trends of properties, ensuring your clients are always ahead in the investment game.

Deciphering Signs of a Hot Property: But how does AI decipher which properties are bound to be lucrative investments? It analyzes various data points like location demographics, local amenities, transportation facilities, upcoming infrastructural developments, and historical price trends. AI scrutinizes these aspects, learning from past patterns, and accurately predicts which properties are bound to witness a surge in value and demand. *Note: predictive analytics is not guaranteed to be accurate.*

Tailoring Investment Strategies: AI doesn't just predict; it strategizes. Based on the insights gleaned from predictive analytics, it helps real estate agents formulate investment strategies for their clients that are rooted in data-driven decisions. It's like tailoring a strategy that's in sync with future market trends, ensuring every investment made today reaps benefits tomorrow. *Note: predictive analytics is not guaranteed to be accurate.*

Unlocking Future Value Today: The real magic of predictive analytics lies in its ability to unlock the future value of a property today. It empowers real estate agents to guide their clients towards properties that are not just valuable now but continue to be so in the future. It's like having a time machine that showcases the future worth of a property, ensuring that every investment made is future-proof. *Note: predictive analytics is not guaranteed to be accurate.*

Creating a Safety Net in Investments: With AI and predictive analytics, real estate agents weave a safety net around their client's investments. It safeguards them from potential market volatilities by steering their investments towards properties that are predicted to not only retain but grow in value, even in fluctuating market conditions. *Note: predictive analytics is not guaranteed to be accurate.*

Sculpting the Path of Growth: For real estate businesses, predictive analytics provides a strategic edge, enabling them to carve a growth path that's aligned with future market trends. It's not just about making sales; it's about ensuring that every property sold or bought through them is a step towards financial growth and stability for their clients. *Note: predictive analytics is not guaranteed to be accurate.*

We have now unveiled how AI, coupled with predictive analytics, transforms real estate agents into futuristic advisers, ensuring every property identified and every investment suggested is a seed sown for future financial prosperity. AI doesn't just work on existing data; it crafts future financial narratives, sculpting a path that ensures every investment made is a tale of success and every property bought is a cornerstone for financial stability and growth.

AI-driven Chatbots for 24/7 Customer Inquiries

Imagine this: It's midnight, and a potential client has stumbled upon your real estate listing. Brimming with curiosity and questions, they want answers right then. Here's where the magic of AI-driven chatbots comes into play, ensuring that every inquiry, regardless of the ticking clock, is acknowledged and attended to, keeping customer interest alive. Let's examine a few of the key benefits of AI-driven chatbots:

The Seamless Blend of Technology and Customer Service: An AI-driven chatbot is like having a friendly, knowledgeable assistant on standby round-the-clock, ready to interact with clients whenever they need assistance. These chatbots, powered by Artificial Intelligence, are equipped to understand and respond to user inquiries, providing them with timely and relevant information, and maintaining a smooth communication channel at all hours.

Bridging Timelines and Time Zones: Real estate is a global industry. A property listed might catch the eye of an international investor or a local first-time buyer. AI-driven chatbots eliminate the limitations imposed by time zones and conventional working hours, ensuring every client, irrespective of their geographic location, receives immediate and accurate responses to their inquiries.

Offering Personalized Experiences: Chatbots aren't just automated response units; they're personalized virtual assistants. They can be programmed to interact with clients in a personalized manner, offering responses tailored to their inquiries and preferences, thus ensuring that every interaction is meaningful and impactful.

Less Waiting, More Engaging: In an industry where timing can be everything, ensuring that client inquiries and concerns are addressed immediately can be a pivotal factor in clinching a deal. Chatbots ensure that no client is left waiting, offering instant responses and maintaining a constant engagement that can significantly enhance customer experience and satisfaction.

Learning and Evolving with Every Interaction: What's enchanting about AI-driven chatbots is their ability to learn and evolve. With every interaction, they absorb new information, understand varied customer inquiries, and continuously enhance their response mechanism, ensuring that with every chat, they become a tad bit more adept at providing impeccable customer service.

An Infusion of Efficiency for Real Estate Agent Operations: For real estate agents, AI-driven chatbots are not just tools of customer interaction; they're efficiency enhancers. By handling routine inquiries and maintaining a constant communication channel with clients, they free up real estate agents to focus on more strategic aspects of client interactions and deal closures.

In a nutshell, AI-driven chatbots seamlessly mesh technology and customer service, offering a platform that's always alive with interaction, information, and engagement, ensuring that every client inquiry is a step towards forging a fruitful relationship and potentially closing a deal.

As we navigate through the forthcoming chapters, we continue to explore the realm of AI in real estate, uncovering facets that continue to revolutionize customer interactions, operational efficiency, and strategic planning within the industry. We will witness how AI, with its multifaceted applications, continues to sculpt a future where technology and human expertise blend to offer services that are not just satisfactory but delightfully exemplary.

Price Predictions and Market Analysis

In the dynamic realm of real estate, property prices exhibit significant variability, akin to complex meteorological shifts. Whether one is an experienced real estate agent or a newcomer to the field, possessing a tool that accurately forecasts future property price trends is invaluable. AI, through its advanced algorithms, provides this capability, offering predictive analytics that navigate the intricacies of property price fluctuations.

Understanding the Core: AI and Predictive Algorithms

In simplest terms, Artificial Intelligence (AI) utilizes data from past property sales, encompassing factors like location, size, features, and market dynamics, to predict future pricing trends. Picture AI as your smart, analytical friend who, after diligently studying years of property price fluctuations, discerns patterns, and gives you a glimpse into **probable** future trends.

Let's walk through a couple of real-world scenarios to understand the profound impact of AI in predicting property price trends.

Example 1: Spotting the Next Property Hotspot with AI

Imagine, you're eyeing two neighborhoods: A and B. While A has steadily been popular, B has seen modest growth. Here's where AI plays its part. By analyzing various factors like emerging businesses, infrastructure developments, and socio-economic indicators in and around these areas, AI can predict that B is on the brink of becoming the next sought-after property hotspot. Thus, helping you make a data-backed decision whether to invest in an under-the-radar area that might yield impressive future returns. (See Appendix A for an example of how this would be created with AI.)

Example 2: Predicting the Best Time to Sell

Consider a property in a bustling urban space. While conventional wisdom might suggest that the prices in such an area will always witness an upward trend, AI digs deeper. By analyzing micro-factors like future development projects, political stability, employment rates, and even potential policy changes, AI might foresee a plateau or even a dip in property prices in the area. This predictive insight could be pivotal in deciding whether to sell now or hold off for a potentially more lucrative future. (See Appendix B for an example of how this would be created with AI.)

The Underpinning Strength: Data

The mighty power of AI in predicting property price trends lies in its capability to sift through voluminous data, discern patterns, and predict possible outcomes. This isn't merely number crunching but an in-depth analysis of data that spans across various facets, offering predictions that are robust, reliable, and remarkably precise.

The Encompassing Impact: Strategic Decision Making

AI doesn't just predict; it empowers. With the insights provided by AI, real estate agents are equipped to make strategic decisions, whether it's identifying potential investment areas, deciding on the ideal listing price, or determining the perfect time to buy or sell.

Empowering Real estate agents with Technological Prowess

AI has emerged as an invaluable partner, particularly for real estate professionals, in maneuvering the ever-changing landscape of the property market. By providing accurate forecasts, AI introduces a level of clarity that is often unattainable with solely human analysis, guaranteeing that each decision in the market is well-thought-out, strategic, and backed by an extensive array of data.

In essence, the adoption of AI in predicting property price trends isn't merely a technological upgrade; it's a strategic enhancement, ensuring real estate agents are not just staying afloat amidst market dynamics but are steering their ship, with confidence, towards lucrative harbors.

AI for Comparative Market Analysis (CMA)

Bridging AI and CMA

The world of real estate is constantly shifting, and with these shifts comes the crucial task of precisely valuing properties, for which Comparative Market Analysis (CMA) is widely utilized. Essentially, the CMA is a strategy that helps real estate professionals determine the value of a property by comparing it to similar properties that have recently been sold. Now, when we blend this strategy with Artificial Intelligence (AI) – a technology that excels at quickly analyzing large sets of data and predicting future trends – the result is a remarkably enhanced approach to pricing properties. This fusion ensures that the pricing strategies developed are not only highly accurate but are also adept at foreseeing and adapting to future market shifts, providing a thoroughly contemporary, agile, and strategically sound framework for real estate valuation.

The Core of AI and CMA

Artificial Intelligence (AI), once deemed a technology of the future, has become a present-day reality, weaving its capabilities into numerous industries, including real estate. In this sector, AI serves as a powerful tool that helps professionals make well-informed decisions by providing advanced data analysis and forecasting abilities, ensuring strategies are both accurate and efficient. On the other side, Comparative Market Analysis (CMA) has long been a conventional method used to determine a property's value. This approach involves comparing the property to similar ones (often referred to as "comps") based on their sale prices and attributes in a particular area. Now, the merger of these two - the innovative analytics of AI and the comparative approach of CMA - stands to revolutionize how property valuation is conducted by combining deep, rapid analysis with tried-and-true valuation techniques.

The Symbiosis: AI in Enhancing CMA

Artificial Intelligence (AI) and Comparative Market Analysis (CMA) come together to forge a tool that is deeply insightful, forward-thinking, and notably quick in real estate valuation. In simpler terms, AI is notable for its ability to quickly sift through and analyze large amounts of data, offering predictions and outcomes based on that data. When applied to CMA, AI enhances it by bringing a precise and futuristic approach to determining property values. It does so by considering a myriad of factors that could influence property value, both directly and indirectly, ensuring that the valuation is not only rooted in the present but also considers potential future shifts and trends in the market. This collaboration aims to ensure that property assessments are as accurate, comprehensive, and strategically sound as possible.

Application in Real-World Scenarios

Here are two scenarios illustrating the practical application of AI in CMA:

Scenario 1: Navigating through Rapidly Changing Urban Markets

Consider a residential property situated in a bustling city where property values are often in flux due to numerous factors like developmental projects, economic fluctuations, and policy changes. Traditional CMA might fall short in encapsulating the entirety of these dynamic influences. Here, AI, with its predictive analysis, comprehensively integrates these variables, providing a robust, dynamic pricing strategy that is both current and anticipatory in nature, ensuring that the property is competitively priced, and anticipates future market shifts.

(For additional information on how you would use AI to address this scenario, visit Appendix C.)

Scenario 2: Analyzing a Suburban Property with Historical Significance

Imagine a suburban property embedded with historical significance, where conventional CMAs might overlook the nuanced impact of this feature on its value. AI can analyze data sets that delve deeper into how historical value impacts property prices within a specific region, considering factors like tourist attraction, cultural importance, and policies related to historical sites. This advanced analytical approach ensures a more contextually accurate, nuanced, and market-relevant valuation, which is particularly pertinent for properties with unique characteristics.

(For additional information on how you could use AI to address this scenario, visit Appendix D.)

AI: A Facilitator for Real estate agents

Incorporating Artificial Intelligence (AI) into the world of real estate and, more specifically, into the Comparative Market Analysis (CMA) process empowers real estate agents with the ability to craft pricing strategies that are profoundly insightful, market-responsive, and strategically aligned with future market trajectories. AI does not merely enhance the CMA; it elevates it to a level where predictive accuracy and analytical depth become intrinsic to property valuation.

Overcoming Challenges and Looking Forward

While AI provides a myriad of advantages, the pathway to its integration in real estate practices is not devoid of challenges. From technological adaptability to ethical considerations in data utilization, real estate agents embarking on this journey must navigate through these aspects mindfully, ensuring that the adoption of AI is not just technologically progressive but also ethically and operationally sound.

Crafting the Future of Real Estate Valuations

The integration of AI into CMAs represents a paradigm shift in real estate valuations and pricing strategy development. It embeds a future-forward, analytical, and strategically nuanced approach into the realm of property valuation, ensuring that pricing strategies devised are not mere reflections of the current market but are insightful anticipations of future trends.

Anticipating Market Shifts Using AI

In the fluctuating world of real estate, being a step ahead is not just an advantage – it's a necessity. The ability to predict and navigate through market shifts can drastically impact strategic planning and decision-making processes. Artificial Intelligence (AI) comes into play as a powerful tool, turning data into actionable insights and equipping real estate agents with the foresight to anticipate market changes.

Changes in the real estate market can happen for various reasons, including changes in the economy, different buying habits of consumers, and new government policies. Things like changes in employment rates, the growth of the economy, and mortgage rates can greatly affect how and when people buy properties. Additionally, the evolving tastes, financial capabilities, and population changes of consumers can be complex and impactful. And of course, new government policies – like changes to taxes, local laws, or investment rules – can have a broad impact on the value of properties and how buyers and sellers interact.

Artificial Intelligence (AI), with its ability to deeply analyze data, explores large amounts of information, carefully studying various factors and how they impact the real estate market. AI doesn't just find but also interprets hidden patterns and trends in both historical and current data, providing a clearer and forward-thinking understanding of possible future market paths.

With the help of AI, predictive analytics becomes a crucial tool for real estate agents, helping them to anticipate, not just react to, changes in the market. Real estate agents can use AI to simplify and clarify large and varied data into clear, actionable advice. For example, by using data about past economic recessions, new consumer trends, and the effects of policy changes, AI can create predictive models that offer solid predictions about possible future market changes and variations.

Moreover, AI helps in planning responses to these expected changes, aligning investment, pricing, and promotional strategies to effectively navigate through the anticipated market situations. For real estate agents, this advance knowledge is key in advising clients, planning investments, and creating flexible, resilient strategies that are not only reactive but are also anticipatively shaped to skillfully navigate through upcoming market changes with sharp precision and foresight. The partnership between technology and real estate professionals enhances the ability to make more educated, strategic, and future-focused decisions in the constantly changing property market.

Mechanisms of AI in Predictive Market Analysis

AI processes extensive data, from property prices to economic indicators, and uses pattern recognition to decode the complexities of market trends. Through machine learning and data analytics, it provides projections, helping real estate agents to strategize by anticipating possible future market conditions.

Example 1: Predicting Urban Development Impact

> **Scenario**: Consider a city planning a massive commercial development, which is expected to enhance job opportunities and consequently, demand for nearby residential properties.

> **AI Application**: By analyzing historical data from similar urban developments, AI can predict potential hikes in property demand and prices in the affected areas, enabling real estate agents to strategize their investments and client advisories accordingly.

Example 2: Navigating Through Economic Downturns

> **Scenario**: Envisage a looming economic downturn that threatens to depress property prices and stall market movement.

> **AI Application**: AI can scrutinize patterns from previous recessions, recognizing trends and enabling real estate agents to strategize based on predicted impacts on property prices, buying behaviors, and investment safety nets.

Future Implications

While AI presents the prospect of enhanced predictive capability and strategic planning, it also brings forth challenges, such as ethical/fair housing considerations in data usage and ensuring technological reliability and accuracy. Balancing AI's capabilities while mitigating its challenges becomes pivotal.

As AI technologies evolve, their predictive accuracies and operational efficiencies are likely to expand. However, this must be balanced with ethical considerations, ensuring that AI operates transparently and without bias.

Harnessing AI to anticipate market shifts propels real estate practices into a future where strategic planning is data-driven, calculated, and proactively aligned with predicted market trends. It's imperative to continually evolve with technological advancements, ensuring ethical and effective AI application in navigating through the multifaceted real estate market landscapes.

The Rise of Virtual Property Tours

This section may revisit familiar concepts in real estate sales. However, to ensure that we are all aligned with the relevant terminology, consider it as a refresher.

Imagine stepping through the front door of a beautifully crafted home, observing the gleaming hardwood floors, the delicate, airy curtains, and the sophisticated furniture. Now, imagine doing all of this without physically being there – welcome to a virtual property tour!

A virtual property tour is an immersive, online experience that allows prospective buyers to navigate through a property and explore its features without leaving their comfort zones. It is developed using panoramic images, videos, 3D models, or a combination of these, enabling users to "walk through" a space at their own pace, scrutinize details, and even visualize different decor or furniture using augmented reality (AR). This technology doesn't just transcend geographical barriers but also molds a new perspective toward property showcasing.

Key Features of Virtual Tours:

3D Walkthroughs: Facilitating a step-by-step journey, allowing viewers to experience the property as if they were there in person.

360-degree Views: Providing a panoramic visual exploration of each room or space.

Interactive Floor Plans: Enabling users to engage with the layout, obtaining an understanding of spatial distribution.

Augmented Reality (AR): Integrating digital elements into the live view, such as virtual furniture staging.

Historical Lens: The Rise and Rise of Virtual Property Tours

To navigate the transition towards understanding the burgeoning popularity of virtual property tours, it's imperative to scroll back a few pages in history.

A Timeline of Evolution:

Pre-2020: Virtual tours were already a part of the real estate technology toolkit, predominantly utilized for high-end properties or clients residing in different geographical locations.

2020 Onwards: A pivotal turn was witnessed as COVID-19 became a global reality, necessitating social distancing and a significant reduction in physical interactions. Virtual tours suddenly transformed from a luxury to a necessity.

Acceleration Due to Unforeseen Circumstances: A Case of the Pandemic

When the global pandemic made its adverse presence felt, it cast a formidable shadow on various sectors, real estate included. Traditional open houses and physical property visits witnessed a staggering decline. Amidst this, virtual property tours emerged as a silver lining, providing a viable solution that adhered to social distancing norms while ensuring that the real estate market persisted.

Work From Home (WFH) Phenomenon and Impact on Real Estate:

The unexpected embrace of work from home (WFH) policies related to the Covid quarantines also played a pivotal role in reshaping real estate marketing and operations.

Global Prospects: The flexibility of remote work enabled individuals to explore properties beyond their immediate locales, fueling the demand for virtual property tours which allowed them to inspect distant and even international properties without travel.

Safety: In a pandemic-stricken world, ensuring safety became paramount. Virtual tours facilitated property exploration without jeopardizing the health of agents and clients.

The Intricacies of Crafting Virtual Tours and Augmented Reality in Real Estate:

Creating a successful virtual property tour and integrating augmented reality (AR) involves a blend of technological prowess and a deep understanding of what a prospective buyer wishes to explore.

Photography and Videography: High-quality, detailed images and videos form the foundation of virtual tours.

Software Proficiency: Leveraging robust software to stitch together visuals, ensuring a smooth, user-friendly virtual experience.

Augmented Reality (AR) Implementation: Utilizing augmented reality (AR) technology to enhance virtual tours, offering viewers the ability to personalize spaces, visualize renovations, or place their own furniture within the virtual space.

Looking Ahead: A Permanent Fixture in Real Estate Marketing

While initially a response to a global crisis, virtual property tours have carved out a permanent space within the real estate industry. These technological advancements not only cater to immediate needs but also lay down a framework for a future where technology and real-world experiences meld seamlessly, broadening the horizons of possibilities in property showcasing.

Sustainable and Accessible

Virtual tours negate the need for numerous physical visits, aligning with global shifts towards more sustainable business practices and offering an inclusive approach that caters to individuals regardless of their location or physical abilities.

The Future is Now

With ongoing advancements in technology, the potential for virtual property tours and Augmented Reality (AR) within the real estate sector is boundless. Expect to witness an increasingly sophisticated and immersive experience that will continue to shape and redefine the way we explore and interact with properties in the digital realm.

As the dynamics of the real estate market continue to evolve in alignment with technological advancements and shifting global realities, embracing virtual property tours and augmented reality is not just an innovative strategy but an essential step forward, fostering connectivity and exploration in a world that is ever-increasingly online.

Enhancing Property Visualization with AR

Augmented Reality (AR) might sound like a concept borrowed from a futuristic sci-fi novel, yet it's an accessible and transformative technology that's reshaping numerous industries, including real estate. But what exactly is it? Simply put, AR overlays digital content on the real-world environment through devices such as smartphones, tablets, or AR glasses, enhancing our perception of reality. In real estate, AR can transform flat, static property listings into immersive, interactive experiences.

Peering Through the AR Lens

Imagine pointing your smartphone towards an empty room and seeing it flourish with stylish furniture, vibrant color schemes, and aesthetic decor – all adjustable to your liking. Imagine being able to see eactly what your future would look like in a new home. This powerful visualization, bridging imagination and reality, is already being delivered through Augmented Reality.

Leveraging AR for Elevated Property Showcasing

How exactly can real estate agents adopt Augmented Reality (AR) technology to bring properties to life and captivate potential buyers?

1. Virtual Staging with AR:

What it Is: Infusing digital elements, like furniture and decor, into an empty space viewed through an AR-enabled device.

Benefits: Agents can showcase the potential of a space without investing in physical staging. Buyers, on the other hand, can visualize properties as livable homes, enhancing emotional engagement.

2. AR Property Tours:

What it Is: A blend of virtual and physical worlds where buyers on-site can utilize AR to explore different layouts, designs, and features.

Benefits: It provides a personalized, immersive experience, allowing clients to "feel" the living space and visualize their future in it.

3. Interactive 3D Property Models:

What it Is: An augmented 3D model of a property that users can explore and interact with through their devices.

Benefits: Enables remote clients to examine properties in detail, promoting engagement and facilitating informed decision-making.

4. AR Navigation and Local Insights:

What it Is: AR applications that, when aimed at a property or location, provide valuable data like property details, price, history, and even neighborhood insights.

Benefits: Enhances physical viewings with a layer of interactive data, delivering a richer, informative experience to buyers.

Integrating AR: A Step-by-Step Guide for Real estate agents

Here's a simplified roadmap for real estate agents aiming to integrate AR into their marketing and showcasing endeavors.

Step 1: Identify the Need

Understand where AR could have the greatest impact – is it in virtual staging, interactive tours, or perhaps local insights or neighborhood tours?

Step 2: Collaborate with Tech Experts

Unless you have an in-house AR expert, collaborate with technological firms or freelancers specializing in AR. It is quite likely that you will not be able to accurately create a true augmented reality without professional support.

Step 3: Content Creation

Develop high-quality, realistic 3D models and images that will be used to create the AR experiences. Remember: the quality of visual content directly influences the overall AR experience amd needs to be consistent with your brand.

Step 4: Develop AR Experiences

Work with your tech team to build immersive, user-friendly AR experiences tailored to your properties and client base.

Step 5: Market and Implement

Integrate AR experiences into your marketing strategy, ensuring they're easily accessible to your target audience through your website, app, or social media.

Navigating Forward: AR as a Staple in Real Estate

AR in real estate has shifted from being a novelty to a valuable tool that can elevate property presentation and client engagement to new heights. By allowing clients to explore, personalize, and intimately experience properties, AR not only enhances property visualization but also brings a client's dream home into the realm of reality, all through the screen of their device.

Remember, the incorporation of AR doesn't just signify keeping up with technological advancements – it represents a commitment to providing enriching, client-centered services, further solidifying your position as a forward-thinking, client-oriented real estate agent.

Benefits for Real estate agents and Clients

The lines between the physical and virtual realms have started to blur. Within the real estate industry, two technologies have emerged as game-changers: Augmented Reality (AR) and Virtual Reality (VR). While both offer unique experiences, their commonality lies in their power to revolutionize property visualization. Let's dive into the tangible benefits they offer to both real estate agents and their clients.

Augmented Reality (AR) – Bridging Real and Digital

AR enhances our real-world environment by overlaying it with digital information and visuals. It's like looking through a window, where digital elements come alive on the existing backdrop.

Benefits for Real estate agents:

Cost-Effective Staging: Instead of investing in physical staging, which can be pricey and time-consuming, AR allows for virtual staging, showcasing properties in various styles and themes.

Interactive Marketing Materials: AR-enabled brochures or business cards can come to life, offering a dynamic way to engage potential clients.

On-Site Enhancements: During physical property visits, AR can overlay valuable information about property features, local schools, or historical sales data.

Benefits for Clients:

Visualize Potential: Clients can see how an empty space might look with their own furniture or a new set of decor, making it easier to imagine themselves in the property.

Personalized Tours: AR gives clients the autonomy to explore properties interactively and to their preference.

Instant Information: Clients can get instant insights about a property's features or the surrounding neighborhood.

Virtual Reality (VR) – Stepping Into a New World

Virtual Reality (VR) offers a fully immersive digital experience, shutting out the physical world entirely. Wearing VR headsets, users can "step into" a digital rendition of a property.

Benefits for Real estate agents:

Remote Property Showcasing: Without geographical constraints, real estate agents can offer virtual property tours to clients anywhere in the world.

Streamlined Showings: Instead of driving from one property to another, real estate agents can showcase multiple properties within a short span, optimizing time and effort.

Enhanced Engagement: The immersive nature of VR can captivate potential buyers, offering a unique, memorable property viewing experience.

Benefits for Clients:

Convenience: Clients can tour properties from the comfort of their own homes, eliminating the need for travel.

Tailored Viewing: VR tours can be curated to highlight features that are most important to individual clients.

Emotionally Engaging: The immersive nature of VR allows clients to "feel" the space, facilitating a deeper emotional connection to potential homes.

Synthesis: AR & VR - A Dual Boon for the Real Estate Realm

Incorporating AR and VR technologies offers dual benefits: it empowers real estate agents to showcase properties in versatile and impactful ways, and it enhances the buying experience for clients, making it more interactive, convenient, and engaging.

ROI for Real estate agents:

The initial investment in AR and VR technologies can yield significant returns. These include:

Saving on physical staging costs.

Attracting a broader, even global, clientele.

Enhancing brand image as tech-savvy and client-centric.

Value for Clients:

Clients are granted a richer, more comprehensive understanding of properties.

They save time and energy in the property search phase.

They're provided with an elevated, modern property viewing experience.

As we delve deeper into the transformative world of technology within real estate, it's evident that the fusion of AR and VR is more than just a trend. It's a progressive step towards redefining property visualization and client engagement. For real estate agents willing to ride the wave of technological advancement, the horizon promises vast potential and unparalleled opportunities.

AI for Paperwork and Legal Procedures

Real estate transactions involve a maze of documents, from property listings to final contracts. While these papers ensure legality and clarity, they can be a formidable task for agents and clients alike. With the advent of Artificial Intelligence (AI), however, this intricate process can be simplified, streamlined, and fortified. Here's a glance at how AI is reshaping real estate paperwork and legal procedures.

AI – Unveiling Its Potential in Documentation

AI doesn't just mimic human intelligence; it processes data at lightning speeds, recognizes patterns, and even learns from experiences, making it an invaluable asset for managing real estate, title, and financial paperwork. Some examples might include:

Automated Data Entry:

What it Is: Using AI-driven software, data from forms can be automatically read and inputted into necessary systems or databases.

Benefits: Saves time and reduces human errors, ensuring data accuracy and reliability.

Document Categorization & Management:

What it Is: AI can automatically classify, tag, and store documents based on their content and significance.

Benefits: Streamlines the organization of files, making it easier for agents to retrieve and track necessary documents. This works especially well in finance (for mortgage brokers) and in the title industry.

Digital Assistants for Document Creation:

What it Is: AI-driven bots can generate required documents based on consumper or agent input provided, using templates and integrating necessary legal clauses.

Benefits: Reduces manual work and ensures standardized, compliant paperwork.

Smart Contract Validation:

What it Is: Using AI, contracts can be scanned for inconsistencies, missing clauses, or potential areas of contention.

Benefits: Offers an added layer of validation, ensuring contracts are foolproof and legally sound.

Automated Due Diligence:

What it Is: AI tools can quickly analyze vast data sets, such as property records, to verify their authenticity and accuracy.

Benefits: Expedited verification process that's more precise, aiding in risk mitigation. Enhanced AI due diligence may also significantly decrease the amount of time it takes to close a transaction.

Fraud Detection:

What it Is: AI algorithms can cross-check documents and identify potential fraud by spotting inconsistencies or anomalies.

Benefits: Increases the security of transactions and protects both agents and clients.

Advantages Beyond Automation

Beyond the direct impact on paperwork, AI brings a host of indirect benefits for the real estate agent and broker. Some examples of these benefits will include:

Efficiency: Speedy processes translate to faster deal closures.

Cost Savings: Reduced manpower and time lead to substantial cost benefits.

Client Satisfaction: A streamlined paperwork process enhances the client experience, positioning the agent or agency as tech-savvy and efficient.

Navigating Potential Pitfalls

While AI brings undeniable advantages, real estate agents should be cautious:

Data Privacy: Ensure that the AI tools used are compliant with data protection regulations.

Over-reliance: AI should complement, not replace, the human touch. Real estate transactions are deeply personal, and a human-centric approach remains crucial.

The Road Ahead: AI as the Modern Real estate agent's Ally

As AI makes deeper inroads into the realm of real estate, its utilization particularly in handling paperwork and legal formalities marks a pivotal transition from labor-intensive manual methods to streamlined, automated operations. This technological advancement empowers real estate professionals to effortlessly traverse the complex labyrinth of legal documentation, freeing them to concentrate on the core aspects of their role: fostering client relationships and successfully finalizing transactions. This shift not only augments efficiency but also reshapes the landscape of real estate transactions.

Predictive Analytics for Assessing Property Legal Risks

In the realm of real estate, due diligence is paramount. Legal risks associated with properties can result in considerable financial setbacks and tarnish reputations. Enter Predictive Analytics powered by Artificial Intelligence (AI) – a revolutionary approach to assessing and mitigating potential legal pitfalls in real estate transactions. This chapter delves into the transformative potential of this synergy.

AI-Powered Predictive Analytics: An Overview

Predictive Analytics harnesses the power of Artificial Intelligence and statistical techniques to delve into historical data, aiming to foresee future occurrences. In the realm of real estate, it scrutinizes historical legal disputes, property litigation cases, and inaccuracies in documentation to estimate the likelihood of legal complications in new deals. Additionally, this analytical method helps real estate professionals and investors make informed decisions by identifying patterns and trends that could impact the success of their transactions. All of this can happen in an enhanced timeframe with reduced errors.

Key Components:

Data Collection: Accumulating vast sets of historical data related to property transactions, legal disputes, zoning changes, and more.

Data Processing: Utilizing AI to process, clean, and organize this data.

Pattern Recognition: AI algorithms identify trends, patterns, and anomalies in the data.

Risk Prediction: Based on the patterns recognized, AI predicts potential legal risks for new or pending property transactions.

Predictive Analytics in Action: Assessing Legal Risks

Title Discrepancies:

> **How it Works:** AI reviews past records of properties with title discrepancies and identifies patterns or commonalities.
>
> **Benefits:** Real estate agents can be alerted to potential title issues before they arise, ensuring smoother transactions.

Zoning and Land Use Violations:

> **How it Works:** By analyzing historical zoning changes and disputes, AI can predict areas or properties that might face future zoning challenges.
>
> **Benefits:** Investors and developers can make informed decisions, reducing the risk of costly legal battles.

Easement and Boundary Disputes:

> **How it Works:** AI examines past easement disputes and boundary issues to forecast potential problem areas.
>
> **Benefits:** Property buyers can be forewarned, allowing for proactive resolution strategies.

Environmental Risks:

> **How it Works:** AI assesses past environmental litigations and

property impacts to highlight areas that might pose future environmental concerns.

Benefits: Ensures compliance with environmental regulations, preventing possible legal repercussions.

Leveraging the Predictions: Strategic Actions

Preemptive Resolutions: By identifying potential legal pitfalls, real estate agents can resolve issues even before they surface, ensuring seamless transactions.

Informed Client Counseling: With the insights garnered, agents can provide better advice to clients about potential risks, bolstering their trust.

Optimized Property Valuation: Properties with lower associated legal risks can be valued higher, while those with potential issues can be re-evaluated.

The Road Ahead: Opportunities and Challenges

While Predictive Analytics offers immense potential, it's crucial to navigate with awareness:

Data Integrity: The accuracy of predictions is as good as the data fed. Regular updates and rigorous data verification are vital.

Legal Frameworks: AI tools should be in compliance with regional data protection regulations.

Human Oversight: AI provides valuable insights, but human expertise is essential to interpret these findings in a nuanced real-world context.

Harnessing the power of AI for Predictive Analytics promises a future where legal risks in real estate can be anticipated and navigated with greater precision. As the realms of AI and real estate further intertwine, the industry stands on the cusp of an era where proactive risk management is not just an advantage, but the norm.

Automating Repetitive Paperwork Tasks

In the detailed world of real estate, paperwork is both essential and time-consuming. From listing forms to final contracts, agents often find themselves buried in documents. Thankfully, the integration of Artificial Intelligence offers a dynamic solution, automating repetitive tasks and liberating agents from the drudgery of paperwork. This chapter unfolds the wonders of AI in revolutionizing real estate documentation.

The Advent of Automated Documentation with AI

AI systems are designed to learn and apply knowledge, making them ideal for repetitive tasks. By understanding patterns and adapting to various documentation needs, AI is positioned to redefine the paperwork landscape in real estate.

Key Implementations:

Document Generation:

How it Works: Input specific property details or client requirements, and AI-driven platforms can generate the necessary paperwork, from listing documents to sales agreements. Automation can be applied to further reduce agent/client input which further reduces errors.

Benefits: Consistency in document creation, reduced manual errors, and time saved.

Form Filling Automation:

How it Works: AI can recognize and fill standard forms using pre-fed data or by extracting relevant information from databases. Gone are the days of errors related to APN number and property address.

Benefits: Speeds up the process, minimizes human intervention, and ensures accuracy.

Document Sorting and Categorization:

How it Works: AI algorithms classify, label, and organize documents based on their content, relevance, and context.

Benefits: Simplifies document retrieval, ensures organized storage and eases the archival process.

Signature and Approval Automation:

How it Works: Digital signatures and AI-driven approval workflows can expedite the sign-off processes. This is not about removing people but reducing errors in the process.

Benefits: Faster document approvals, secure and verifiable signatures, and a more streamlined communication process.

Document Analysis and Error Detection:

How it Works: AI scans documents for inconsistencies, missing details, or potential errors by comparing them against set templates or criteria.

Benefits: Enhances the quality of documentation and ensures legal and procedural compliance.

Elevating the Real Estate Experience

The automation of repetitive paperwork tasks offers multi-dimensional benefits:

Efficiency: By cutting down manual documentation time, deals progress faster.

Accuracy: Automated processes reduce the risk of human errors, ensuring the reliability of the documents.

Client Satisfaction: With quicker turnarounds and transparent processes, clients experience smoother transactions.

Potential Challenges and Solutions

While AI offers unparalleled advantages, it's crucial to be aware of potential pitfalls:

Over-reliance: Remember, AI aids the process but does not replace the need for human oversight. Regularly reviewing automated documents ensures quality.

Data Security: Use AI platforms that adhere to robust security protocols to protect sensitive client data.

Continuous Learning: As real estate laws and protocols evolve, it's essential to update and train the AI systems accordingly.

Glimpsing the Future: AI as the Ultimate Assistant

As we wrap up this chapter, let's acknowledge the growing impact of AI in simplifying real estate paperwork. The transition away from manual, time-consuming tasks is becoming more apparent. This shift allows real estate agents to concentrate on their core strengths: cultivating client relationships, comprehending individual needs, and finalizing property deals effectively.

Embracing AI's capability for automation marks a significant step forward for the real estate sector. It brings about increased efficiency and accuracy. With routine tasks now handled by AI, agents can redirect their focus towards building stronger client connections and advancing their professional development. This practical application of AI is not just about adopting new technology; it's about enhancing the effectiveness and satisfaction in the real estate profession.

Client Relationship Management (CRM) and AI

Understanding AI-Driven CRMs: A traditional CRM system is a software tool used by businesses, including real estate agencies, to manage interactions with clients and potential customers. It helps in organizing information like contact details, communication history, and client preferences. These systems enable agents to track leads, schedule appointments, and follow up with clients. They are primarily manual in operation, requiring agents to input and update data regularly.

An AI-driven CRM system, on the other hand, incorporates artificial intelligence to enhance these capabilities. It goes beyond simply storing information by analyzing data to provide insights and predictions. The AI-driven CRM system does more than organize client information; it actively assists in decision-making, saves time through automation, and provides valuable insights to enhance the productivity and effectiveness of real estate agents.

Deciphering AI-Driven CRMs

At its core, a CRM system is a digital toolkit designed to manage interactions with current and potential clients. When infused with AI, a CRM evolves from being a passive system of record-keeping to an active assistant, offering predictive insights, automating tasks, and personalizing interactions.

Key Features:

Predictive Analytics: Anticipates client needs based on historical data and market trends.

Automation: Streamlines tasks like follow-ups, appointment scheduling, and information updates.

Personalization: Curates client-specific content and communication based on preferences and interaction history.

Real Estate Transformed: Examples of AI-Driven CRM

1. Personalized Property Recommendations:

Scenario: Consider a family looking for a spacious home with proximity to schools and parks. Rather than manually sorting through listings, an AI-driven CRM analyzes the family's preferences, previous searches, and interaction history.

How AI Helps: Using deep learning algorithms, the CRM identifies and recommends properties that closely match the family's criteria, enhancing the chances of a swift and satisfactory deal closure.

Outcome: The real estate agent provides highly relevant property options, impressing the family with their efficiency and understanding of their needs.

2. Automated Client Engagement and Follow-ups:

Scenario: After a successful property viewing, a potential buyer has a few concerns that need addressing. Given a real estate agent's busy schedule, there's a risk of the buyer feeling neglected if their concerns aren't promptly addressed.

How AI Helps: The CRM, recognizing the buyer's engagement level and the potential value of the deal, automatically sends a personalized follow-up email or message, assuring them that their concerns will be addressed shortly. It can even schedule a call or a meeting, ensuring the lead remains warm.

Outcome: The buyer feels valued and heard, increasing trust and loyalty towards the real estate agent, even in their absence.

Leveraging AI-Driven CRMs: Steps for Real estate agents

Data Integration: Feed the CRM with comprehensive client data. The richer the data, the better the AI can operate.

Training: While AI-driven CRMs are intuitive, take the time to understand its features to maximize benefits.

Feedback Loop: Regularly review and provide feedback on the CRM's recommendations and actions to refine its algorithms.

The Future is Personal

In real estate, a one-size-fits-all approach is obsolete. Clients demand personalized experiences and rapid responses. An AI-driven CRM isn't just a tool; it's a game-changer, empowering real estate agents to foster deeper, more meaningful relationships with their clients. By anticipating needs, automating repetitive tasks, and delivering tailored experiences, AI-driven CRMs are setting the gold standard in client relationship management for the real estate industry.

See Appendix E for some examples of companies currently using AI in the CRM.

Personalizing Client Interactions Using AI Insights

In real estate sales, building and nurturing client relationships is at the core of success. While traditional approaches have their merits, the incorporation of Artificial Intelligence (AI) into Client Relationship Management (CRM) has revolutionized how real estate agents can create deeply personalized interactions. Let's delve into how AI insights enable real estate agents to tailor their client interactions, enhancing both client satisfaction and business outcomes.

The AI Advantage: Beyond Generic Interactions

In an era where consumers are inundated with information, personalization stands out. AI, with its capability to process vast amounts of data quickly and draw actionable insights, empowers real estate agents to move beyond generic interactions.

Key Ways AI Drives Personalization:

Behavioral Analysis: By studying a client's online activity, such as property views, search patterns, and response times, AI can predict what a client is truly looking for, even if they haven't explicitly stated it. Even if they don't know what they're looking for. When integrated correctly, AI can even help analyze viewing patterns to create takeaways from how much interaction users spend with a photo or photos.

Sentiment Analysis: Through analyzing textual communication, AI can gauge a client's emotional tone, helping real estate agents adjust their approach accordingly. For instance, if a client seems hesitant, the real estate agent might offer additional reassurance or information.

Historical Data Utilization: AI can utilize past transaction data to understand client preferences and predict future ones, ensuring that property suggestions are always relevant.

Applying AI Insights: Strategies for Personalized Interactions

Tailored Property Suggestions: Instead of overwhelming clients with countless listings, AI insights enable real estate agents to present a curated list that aligns closely with a client's preferences and behavioral patterns. AI will even learn what they like without them having to express those preferences.

Customized Communication: AI can segment clients based on their preferences and engagement levels. This allows for communication strategies, such as personalized emails or messages, tailored to each segment's unique needs. AI can even change tone and content for multiple parties looking at the same properties.

Predictive Engagement: By anticipating a client's queries or concerns based on their interaction patterns, real estate agents can proactively address them, enhancing client trust.

Chatbots with a Personal Touch: Modern AI-driven chatbots can provide instant responses to client queries. But more importantly, they can be programmed to interact in a manner that mirrors the brand's voice and ethos, ensuring consistency in client interactions.

Real-world Example: The Power of Personalization

Consider the scenario of Jane, a prospective homebuyer in the market for a new home. Her real estate agent, utilizing the capabilities of an AI-driven CRM system, is able to accurately assess and understand her unique preferences and requirements. The CRM system, through its advanced analytics, discerns that Jane has a young family, which influences her interest in properties situated in close proximity to schools. Additionally, it notes her penchant for homes that boast attractive gardens. Armed with this tailored information, the agent strategically filters and sends Jane a carefully curated selection of listings. These listings are not just any homes, but those specifically located near reputable schools and featuring charming gardens. This targeted approach not only streamlines Jane's search for the perfect home but also significantly enhances the chances of her arriving at a decision swiftly. This efficiency is mutually beneficial, facilitating a smoother and more satisfying property buying experience.

Concluding Thoughts: The Future is Personal

As the real estate landscape becomes increasingly competitive, the ability to offer extremely personalized experiences can set a real estate agent apart. AI-driven insights provide the tools needed to understand clients on a deeper level, allowing for interactions that resonate and foster lasting relationships.

Predicting Client Needs and Preferences

In the fast-paced realm of real estate, understanding your client is paramount. The ability to anticipate their needs and preferences can be the key difference between a successful transaction and a missed opportunity.

With the integration of Artificial Intelligence in Client Relationship Management (CRM), real estate agents now have a powerful ally in deciphering and predicting client behaviors and desires.

Let's explore how.

The Power of Prediction: Why It Matters

Before delving into the "how," it's crucial to understand the "why." Predicting client needs:

> **Enhances Client Satisfaction:** Clients feel valued when their preferences are anticipated, leading to higher satisfaction levels.
>
> **Increases Efficiency:** Time isn't wasted on properties or deals that aren't a likely fit. The agent can be more responsive by using highly targeted information to find the right property.
>
> **Boosts Sales:** Accurately meeting client needs leads to quicker transaction closures.

The AI Mechanism: Making Sense of Data

At the core of AI's predictive prowess is its ability to analyze vast amounts of data, identify patterns, and draw actionable insights from them. Here's how it's done:

Data Aggregation: AI systems gather data from various sources, such as previous transaction histories, client interactions, online activity, and even social media behavior.

Pattern Recognition: By analyzing this data, AI identifies patterns. For instance, it might recognize that a client frequently searches for properties with home offices, revealing a preference.

Predictive Analysis: Using the patterns it identifies, AI can predict future behavior or preferences. For example, if a client often inquires about energy-efficient homes, the system might predict an interest in properties with solar panels.

Implementing Predictive Insights: Strategies for Real estate agents

Personalized Property Recommendations: Use AI's insights to offer listings that align closely with a client's predicted preferences. If a client often looks at homes near parks, ensure they know about new listings in such areas.

Proactive Communication: If AI predicts a client might be interested in a particular upcoming neighborhood or property type, real estate agents can proactively reach out with relevant information.

Adjusting Marketing Strategies: If AI identifies that a significant portion of clients show interest in, say, eco-friendly properties, real estate agents can adjust their marketing to highlight such features.

Real-world Example: Predictive Power in Action

John, a recent retiree, has been browsing properties. While he hasn't explicitly mentioned it, his online behavior, combined with data from similar demographics, reveals a preference for quieter neighborhoods, proximity to healthcare facilities, and low-maintenance properties. Using these AI-driven insights, his real estate agent can provide a curated list of properties, increasing the likelihood of a swift and satisfying transaction.

The Road Ahead: Continual Learning and Refinement

One of the most potent aspects of AI is its ability to learn continually. As it processes more data, its predictions become even more accurate, ensuring that real estate agents are always one step ahead in understanding and meeting client needs.

In conclusion, in the dynamic world of real estate, the ability to predict client needs and preferences isn't just a luxury—it's a necessity. And with AI-powered CRM, real estate agents are equipped to navigate this challenge with unparalleled precision. Embracing these technologies and understanding their potential can elevate the client experience, making the home-buying or selling journey seamless, efficient, and remarkably personalized. The future of real estate lies in leveraging these predictive insights to create memorable client interactions and foster lasting relationships.

Improving Marketing and Outreach with AI

Targeted Advertising Using AI Analytics: In the digital age, where data is abundant, it's no longer about casting the widest net, but about casting the smartest one. Artificial Intelligence is revolutionizing the world of real estate marketing by refining and personalizing advertising strategies like never before. This chapter will illuminate how AI and analytics can empower real estate agents to engage in precision-targeted advertising, ensuring that marketing efforts are both efficient (potentially saving money) and effective.

Understanding the Power of Data

Before delving into AI's role, it's essential to grasp the significance of data in marketing. Every online interaction, from clicking on a property listing to liking a social media post, generates data. This data, when analyzed, provides valuable insights into potential clients' preferences, habits, and buying behaviors.

AI: Beyond Traditional Analytics

Traditional analytics provides a broad understanding of your audience, while AI dives deeper. It identifies patterns, predicts future behaviors, and even prescribes marketing strategies. For instance, if a user frequently visits listings of beachfront properties, AI can predict their preference and suggest similar properties in their advertisements.

At its core, AI analytics uses machine learning algorithms to analyze vast amounts of data, drawing patterns and insights that would be impossible—or at least extremely time-consuming—for humans to discern. For real estate agents, this means a goldmine of information about potential clients: their preferences, behaviors, and even purchasing intentions.

Personalized Ad Content

By examining data such as online browsing habits, social media interactions, and even responses to previous ad campaigns, AI can segment potential clients into distinct categories. Whether they're first-time buyers, looking for luxury properties, or downsizing, AI can pinpoint these groups with surprising accuracy.

AI analytics allows for the creation of personalized advertisements. Instead of generic property ads, a potential client can receive an advertisement tailored to their preferences, increasing the likelihood of engagement. For example, once these segments are identified, real estate agents can craft personalized ads for each group. For instance, first-time buyers might receive ads featuring affordable starter homes, while those identified as luxury seekers might see ads for high-end properties with upscale amenities. This increases the ad's relevance and, consequently, its effectiveness.

Optimal Timing and Platform Placement

Not only can AI determine what content to show, but it can also decide when and where. By analyzing user activity, AI can ascertain the best time to display an ad and on which platform (e.g., Facebook, Instagram, Google). This ensures that advertisements reach potential clients when they're most receptive.

Beyond current campaigns, AI analytics can predict future market trends. For instance, if AI detects an increased interest in a particular neighborhood or a specific type of property, real estate agents can adjust their advertising strategies accordingly, staying ahead of the curve.

Budget Optimization

AI analytics provides insights on which advertisements are performing best, allowing real estate agents to allocate their budget more effectively. Instead of spending equally across different ad campaigns, funds can be channeled into the most promising ones, maximizing return on investment. AI can identify which platforms yield the best results for specific segments, ensuring that every dollar spent is used effectively.

Continuous Learning and Adaptation

The beauty of AI analytics lies in its ability to learn and refine its strategies continually. As it receives more data from ad campaigns, it adjusts its algorithms, ensuring that advertising strategies are always at their most effective, eliminating the guesswork often associated with traditional advertising methods.

Looking Ahead: The Future of AI in Advertising

As AI technology advances, we can expect even more nuanced advertising capabilities. Soon, virtual property tours might be personalized in real-time based on viewer reactions, or chatbots might provide instant property suggestions based on a potential buyer's conversation.

Where We Are Now

The integration of AI analytics in real estate advertising isn't just a trend; it's the future. By enabling precision-targeted advertising, AI ensures that real estate agents can reach the right people with the right message at the right time. By embracing AI analytics in advertising, real estate agents are setting themselves up for success in an increasingly digital and data-driven world. The result? More engaged clients.

Chatbots for Lead Generation

As the digital age progresses, the ways we communicate and engage with potential clients evolve. In the realm of real estate, the rise of chatbots presents an exciting opportunity for agents to generate leads more efficiently and effectively. But what exactly is a chatbot, and how can it revolutionize lead generation for real estate agents? Let's find out.

What is a Chatbot?

A chatbot is a software application designed to simulate human conversation. It can communicate with users either through text or voice interactions, depending on its design and platform. Chatbots can operate via a variety of digital channels, including websites, social media platforms, messaging apps, or even standalone apps. Powered by AI and machine learning, chatbots can provide instant responses to user queries, streamline interactions, and even gather essential data from potential leads. Don't worry, the chatbots of tomorrow are nothing like what you've seen in the past.

Here are some key aspects of chatbots:

> **Interactivity:** Chatbots are designed to interact with users in real-time, providing instant responses to user queries.
>
> **Automation:** Most chatbots are programmed to respond automatically based on a set of predetermined rules or scripts. They can handle multiple user interactions simultaneously.
>
> **AI-Powered:** Advanced chatbots utilize Artificial Intelligence (AI) and Natural Language Processing (NLP) to understand user intent and generate responses. This allows them to learn from user interactions and improve over time.
>
> **Use Cases:** Chatbots have diverse applications, ranging from customer support, ordering services, providing information, to more complex tasks like assisting with tech troubleshooting or offering personalized product recommendations.

24/7 Engagement

The real estate market doesn't sleep, and neither should your lead generation efforts. With chatbots, real estate agents have a tool that's available around the clock. Whether a potential client has a question at 3 pm or 3 am, chatbots ensure they're attended to immediately, increasing the chances of engagement and capturing a lead.

Qualifying Leads

Not every visitor to your website or digital platform will be a genuine lead. Chatbots can help filter out serious inquiries from casual ones. By asking a series of predefined questions, chatbots can gauge the intent and readiness of a potential client, ensuring that real estate agents focus their efforts on the most promising leads.

Personalized Interactions

With AI capabilities, chatbots can provide personalized responses based on user behavior and preferences. For instance, if a visitor shows interest in luxury properties, the chatbot can tailor its questions and suggestions towards high-end listings, creating a more tailored and engaging user experience.

Efficient Data Collection

One of the primary benefits of chatbots is their ability to gather and store information seamlessly. As potential clients interact with the chatbot, agents receive valuable data about their preferences, budget, location interests, and more. The AI systems, and consquently the agents, are continually learning about the needs and wants of their clients. This data can be instrumental in crafting targeted marketing campaigns in the future.

Integrating with Other Digital Platforms

Chatbots can easily integrate with other digital tools and platforms, such as CRM systems. This integration ensures that all lead data collected by the chatbot is automatically updated in the CRM, streamlining the lead management process for real estate agents.

Building Trust Through Instant Responses

In the digital age, users expect quick answers. Chatbots can provide instant responses to common questions about properties, pricing, neighborhood details, and more. This rapid response not only aids lead generation but also fosters trust and reliability in the eyes of potential clients.

Conclusion

In an industry as competitive as real estate, staying ahead of the curve is crucial. Chatbots offer a modern, efficient, and highly effective way to engage with potential clients, gather valuable data, and generate quality leads. As AI technology continues to advance, the capabilities of chatbots will only grow, making them an indispensable tool for real estate agents looking to thrive in the digital era.

(For additional information on chatbots specific to real estate, see Appendix F.)

Social Media Insights and Optimization Using AI

For real estate agents seeking to broaden their influence and forge enduring connections with prospective clients, social media proves to be an essential resource. Yet, mastering the extensive and intricate world of social media is often a challenging endeavor. This is where Artificial Intelligence steps in as a transformative asset. It unravels the intricacies of social media patterns, tendencies, and algorithms, thereby enhancing the efficacy of marketing strategies.

Understanding AI-Powered Social Media Insights

Data Analysis at Scale: AI algorithms can process vast amounts of data much faster than a human can. This means analyzing the behavior, preferences, and engagement patterns of thousands, even millions, of social media users in real time.

Sentiment Analysis: By scanning comments, shares, and likes, AI can gauge the general sentiment toward a property listing or a real estate agency. This can help agents understand what's resonating with their audience and what's not.

Predictive Analytics: Using historical data, AI can predict future trends. For instance, if there's a rising interest in waterfront properties, real estate agents can capitalize on this trend early.

Optimization Social Media Using AI

Content Personalization: AI can segment audiences based on their behavior and preferences, allowing agents to tailor content that resonates with specific groups. For example, first-time homebuyers might receive content about the home-buying process, while luxury property enthusiasts might see listings of upscale homes.

Optimal Posting Times: AI can determine when a target audience is most active, recommending the best times to post to maximize engagement and reach.

Ad Targeting: AI can help refine ad campaigns by identifying the characteristics of users most likely to engage or convert. This ensures a higher return on ad spend.

Visual Recognition

With AI, it's possible to scan images and videos to determine which visual elements capture the most attention. Real estate agents can use this insight to share property photos and videos that are more likely to resonate with viewers.

Real-World Applications

Chatbots for Engagement: On social media platforms, AI-driven chatbots can instantly respond to queries about property listings, guiding potential clients through the initial stages of inquiry.

Trend Forecasting: Agents can stay ahead of the curve by using AI to detect emerging trends in real estate, from property features to emerging neighborhoods, and adjust their marketing strategies accordingly.

The intersection of AI and social media offers a goldmine of opportunities for real estate agents. By harnessing these insights and optimization techniques, agents can not only enhance their online presence but also create meaningful, personalized interactions that can convert followers into clients.

Remember, while AI provides valuable tools for understanding and engagement, the human touch in real estate remains irreplaceable. Use AI as a tool to enhance, not replace, personal interactions.

Challenges and Ethical Considerations

The Importance of Data Privacy: In the age of Artificial Intelligence (AI), the real estate industry, like many others, has seen a transformational shift. AI algorithms rely on vast amounts of data to make accurate predictions, personalize experiences, and automate tasks. However, as we integrate AI deeper into our services, the issue of data privacy surfaces as a paramount concern. This chapter aims to shed light on the importance of data privacy in AI, specifically in the realm of real estate.

The Data Dependency of AI

Before delving into privacy concerns, it's vital to understand the nature of AI. Artificial Intelligence, especially machine learning models, thrives on data. The more data the individual model receives, the better it performs. For real estate agents, this data can range from clients' personal information, preferences, financial details, to their interactions with various listings. By feeding this data into AI systems, agents can better predict client needs, automate tasks, and optimize marketing strategies.

Artificial Intelligence, at its core, is designed to mimic human cognitive functions, and this imitation is fueled primarily by data. To truly grasp AI's profound reliance on data, one must first understand its different facets and mechanisms.

> **Nature of Machine Learning**: Machine Learning, which falls under the umbrella of Artificial Intelligence, is centered around the concept of training algorithms to identify patterns and make informed decisions from the data provided. This process is comparable to the way a child learns: just as a child develops a better understanding and predictive ability with an increasing number of examples (data), an algorithm (akin to the child) improves its accuracy and predictive capabilities as it processes more data.

Quality vs. Quantity: While AI thrives on large volumes of data, the quality of this data is equally critical. For instance, inaccurate or biased data can lead AI systems to make flawed or prejudiced decisions. In the realm of real estate, ensuring the data's accuracy, from property valuations to client feedback, is paramount.

It is also important to note that AI only has access that it's programmed to have access to and can occasionally produce responses that are not perfectly accurate or up-to-date. In cases where it doesn't have specific information, it might generate responses based on related knowledge or context available in its training.

Diverse Data for Holistic Analysis: For real estate professionals, the spectrum of data can be vast. It's not just about a client's name or contact details. Their home preferences, past property visits, feedback on showings, financial capability, and even their response time to communications can provide invaluable insights. This wide-ranging data allows AI to create a holistic client profile, leading to more personalized and effective services.

Integrating demographic data with a real estate Customer Relationship Management (CRM) system, powered by AI, can significantly enhance the home search experience for consumers. Here's how it works:

Understanding Consumer Preferences: AI can analyze demographic data such as age, family size, income level, and lifestyle preferences. This information helps in creating detailed consumer profiles.

Tailored Property Recommendations: Based on these profiles, the AI system can identify properties that align with the specific needs and preferences of different demographic groups. For example, young families might prefer homes near schools and parks, while retirees might look for quieter neighborhoods with easy access to healthcare facilities.

Market Trend Analysis: AI can use demographic data to identify emerging trends in different areas. For instance, if a neighborhood is becoming popular among young professionals, the system can highlight properties in that area to buyers fitting that demographic.

Price Range Suitability: By understanding the financial demographics of consumers, AI-driven CRMs can suggest properties that fit within their budget, making the search more efficient.

Cultural and Community Preferences: Demographic data can also include cultural and community preferences, enabling the CRM to suggest neighborhoods that align with the buyer's lifestyle and community desires.

Personalized Communication: Using demographic insights, the CRM can personalize communication with clients. It can send relevant information, listings, and updates that resonate with the consumer's specific demographic.

Predictive Analytics: AI can predict future market trends based on demographic shifts, helping consumers make informed decisions about where to buy a property.

By integrating demographic data, AI enhances the capability of real estate CRMs, making them not just tools for managing client information, but proactive assistants in the home-buying process.

Continuous Learning: Unlike traditional software that remains static, AI systems evolve. As they are exposed to new data, they adapt, refine their predictions, and improve their performance. For real estate agents, this means that an AI tool, over time, can become increasingly adept at understanding local real estate market trends or specific client preferences.

Empowering Real estate agents: The crux of integrating AI in real estate is to empower agents to be more efficient and informed. With the insights derived from the analyzed data, real estate agents can fine-tune their marketing campaigns, anticipate client concerns, and streamline operational tasks. In essence, AI becomes an invaluable assistant, continuously learning from the influx of data and providing actionable insights.

Why Data Privacy Matters

Trust and Reputation: For real estate agents, trust is the foundation of their relationship with clients. If clients fear their data might be misused or mishandled, it could tarnish the agent's reputation and hurt their business.

Legal Implications: Various global regions, including the European Union with its General Data Protection Regulation (GDPR), have established stringent regulations for data protection. Failure to adhere to these regulations can result in substantial financial penalties and legal challenges.

Ethical Responsibility: Apart from regulatory requirements, there exists a moral responsibility to honor and safeguard clients' personal data. The consequences of data breaches or misuse can be deeply personal, ranging from financial damage to significant emotional turmoil for individuals.

Data Privacy Challenges in AI for Real estate agents

Data Collection: AI systems require vast amounts of data, which means there's a temptation to collect more than what's necessary. It's crucial to ensure that only essential data is collected, and clients are informed about its intended use.

Data Storage and Security: Once collected, this data needs to be stored securely. Real estate agents must invest in robust security measures, ensuring data is encrypted and protected from breaches.

Third-party Integrations: Many AI solutions in the real estate industry come from third-party providers. It's essential to scrutinize their data handling and privacy policies.

Best Practices for Ensuring Data Privacy in AI

Transparency with Clients: Always inform clients about the data being collected, how it'll be used, and the measures in place to protect it.

Regular Audits: Periodically assess and update security measures. Regularly audit stored data and delete any information that's no longer needed.

Training and Awareness: Ensure that everyone in the agency understands the importance of data privacy and is trained on best practices.

Collaborate with Experts: Consider collaborating with data privacy and security experts to keep updated with the latest threats and protection measures.

The convergence of AI and real estate offers exciting opportunities for growth, efficiency, and enhanced client experiences. However, with great power comes great responsibility. By prioritizing data privacy, real estate professionals can ensure they harness the benefits of AI without compromising the trust and security of clients.

See what a privacy policy with AI language might look like in Appendix G.

Bias in AI Algorithms and its Implications

Understanding Algorithmic Bias

At its core, AI learns from data. If the data it learns from is biased, the AI can, and likely will, demonstrate biased behaviors. For real estate agents using AI, it's crucial to understand the source and nature of these biases, as they can inadvertently perpetuate stereotypes or lead to unfair practices.

1. What is Algorithmic Bias?

Algorithmic bias refers to situations where AI systems display prejudiced results due to flaws in their design, data, or underlying algorithms. These biases can arise from the data the AI was trained on, the way the algorithm was designed, or even from the people who designed them.

2. Sources of Bias

>**Training Data Bias**: If the data used to train an AI system has inherent biases, the AI will adopt these. For instance, if an AI system for property valuations is trained primarily on properties from affluent areas, it may undervalue properties in less affluent areas.

>**Design Bias**: Sometimes, the way an algorithm is designed can introduce bias. If not crafted carefully, the algorithm might emphasize certain data points over others, leading to skewed results.

Human Bias: The beliefs or prejudices of those developing AI systems can inadvertently get encoded into the system.

3. Implications in Real Estate

Fairness Concerns: An AI tool used for credit checks might favor certain demographics over others if the data it was trained on is skewed, leading to unfair loan approvals or denials.

Property Valuations: Biased AI could undervalue properties in less-known or less-popular areas, affecting sellers negatively.

Recommendation Systems: If a property recommendation system is biased, clients might only see properties in certain areas, missing out on potential homes they might love.

4. Addressing and Preventing Bias

Diverse Data Sets: Ensure that the data used to train AI systems is representative of all demographics and regions.

Regular Audits: Conduct periodic reviews of AI recommendations and outputs to identify any signs of bias.

Open Feedback Loop: Allow users to flag questionable AI decisions, creating a feedback mechanism to continuously improve the system.

Ethical Guidelines: Establish guidelines to ensure that AI development and deployment prioritize fairness and inclusivity.

5. The Way Forward

Awareness is the first step. By understanding the potential pitfalls and biases AI can introduce, real estate agents can make informed decisions about the tools they use. By prioritizing fairness, transparency, and ethics in AI, the real estate industry can leverage the power of advanced technology without compromising on integrity.

Remember, while AI can be a powerful tool in the realm of real estate, it's essential to use it responsibly. Just as a real estate agent would ensure the authenticity of a property's documents, they must also ensure the fairness of the AI tools they employ.

The Human Touch vs. AI: Striking a Balance

While AI offers transformative capabilities in the real estate sector, there's an irreplaceable value in human touch. For real estate agents, striking a balance between utilizing AI's efficiency and preserving the personal connection is essential for lasting success.

Why Balance Matters

Real estate isn't just about buying or selling properties; it's about helping individuals find a place to call home, or helping them transition to new phases of their lives. This journey is deeply personal and often emotional. While AI can simplify many processes, it can't replicate the empathy, understanding, and trust that a human agent brings to the table.

Symbiotic Relationship: AI and Real estate agents

Example 1 - **Lead Generation**:

AI: Chatbots on a real estate agent's website can interact with visitors, collecting preliminary information and gauging interest.

Human Touch: Following up on those leads with a personal call or meeting to understand their unique requirements and aspirations better.

Example 2 - **Property Recommendations**:

AI: Analyzing vast databases to shortlist properties based on specific criteria.

Human Touch: Visiting properties with clients, understanding their reactions, and adjusting recommendations accordingly.

Example 3 - **Price Negotiations**:

AI: Predicting fair market values using data analytics.

Human Touch: Engaging in negotiations, considering not only the price but the emotions, motivations, and urgencies of all parties involved.

Preserving Personal Connection

Leverage AI for Mundane Tasks: Use AI for repetitive tasks like scheduling, sending reminders, or sorting through listings. This frees up more time to engage with clients on a deeper level.

Emphasize Communication: AI can provide insights, but it's up to the real estate agent to communicate these insights effectively, addressing any concerns and ensuring clients feel heard.

Continued Education and Training: As AI tools evolve, real estate agents should stay updated, ensuring they can maximize these tools while still maintaining their unique human touch.

Challenges in Balancing

Over-reliance on AI: With the allure of efficiency, there's a danger of becoming too dependent on AI, potentially sidelining human judgment and intuition.

Loss of Personalization: While AI can personalize recommendations based on data, it lacks the nuance of human understanding, potentially making clients feel like just another number.

The Way Forward

To navigate the evolving landscape of real estate with AI, professionals must remember the core of their role: facilitating deeply personal decisions for their clients. AI should be a tool, not a replacement. By leveraging the best of both worlds, real estate agents can offer unmatched service, combining efficiency with genuine care and understanding.

Remember, while the capabilities of AI are ever-expanding, the essence of real estate remains in understanding and serving people. As real estate agents harness the power of AI, they must also strengthen their commitment to personal connection and trust.

Getting Started with AI

Identifying Your Business Needs: Incorporating AI into real estate practices can seem both overwhelming and daunting, but it doesn't have to be. One of the first steps in making the leap into AI-enhanced real estate is identifying your specific business needs and then diving in to find where AI might benefit you. By doing so, you can choose the most appropriate AI tools that will truly benefit your practice.

1. Understand What AI Can Offer

Before determining where AI fits into your business strategy, it's crucial to understand its potential. AI can provide data analysis, customer behavior predictions, automate repetitive tasks, enhance customer service with chatbots, and improve the qualit of your business while reducing your workload. Having a basic understanding of AI's capabilities will help you pinpoint areas of your business that can benefit from AI now and in the future.

2. Assess Your Current Operations

Data Collection: Are you manually gathering data about properties, markets, or customer preferences? AI can automate this, ensuring you have the most up-to-date information at your fingertips.

Client Interactions: Are you spending hours responding to client queries? Chatbots and AI-driven customer service tools can handle routine inquiries, freeing up your time for more complex interactions.

Marketing Efforts: Examine your current marketing strategies. AI can analyze vast amounts of data to tailor marketing campaigns, target potential leads, and predict market trends.

3. Identify Pain Points

What aspects of your business take up the most time? Where are the inefficiencies? Maybe it's in lead generation, client follow-ups, or data entry. Recognizing these pain points will guide you to specific AI solutions designed to address those challenges.

4. Consider Your Budget

AI tools vary in cost, from affordable off-the-shelf solutions to more bespoke systems. AI is rarely a "one and done" project; make sure you budget for future requirements, maintenance, and training. Understand your budgetary constraints and keep them in mind when exploring available AI tools.

5. Engage with Your Team

Your team, especially those on the ground, often have firsthand knowledge of operational challenges. Regular brainstorming sessions can offer insights into areas where AI can assist. Their feedback can be invaluable in shaping your AI strategy.

6. Prioritize Needs

Once you've listed potential areas for AI implementation, prioritize them. Start with what will provide the most immediate value or address the most pressing challenge. This ensures a smoother transition into AI integration and allows you to experience benefits sooner.

7. Stay Updated

The world of AI is ever-evolving. Regularly attending workshops, seminars, or online courses will keep you updated about the latest developments. This ensures that you're always leveraging the most recent and effective tools for your business.

Identifying business needs relevant to AI doesn't have to be a complicated process. By assessing your current operations, understanding the capabilities of AI, and keeping the lines of communication open with your team, you can pinpoint exactly where AI will be most beneficial. Remember, the goal isn't just to adopt new technology, but to streamline operations, enhance client interactions, and ultimately, drive your business forward.

List of Popular AI Tools for Real estate agents

1. Chatbots and Virtual Assistants

Structurely's Holmes: This is an AI-driven chatbot designed specifically for real estate. It interacts with leads 24/7, ensuring immediate responses.

Offrs: Uses AI to predict which homeowners are likely to sell their houses soon, helping real estate agents target their efforts.

2. Predictive Analytics

Revaluate: Helps agents predict who in their database is most likely to move within the next three to six months.

Likely.ai: Utilizes AI to predict when homeowners are likely to sell, allowing agents to focus on those leads.

3. Virtual Tours and Augmented Reality

Matterport: Offers 3D virtual home tours, enabling clients to explore properties without physically visiting them.

Zillow 3D Home: An AI-powered mobile platform that allows real estate agents to capture 3D tours of homes using their smartphones.

BoxBrownie.com: A profesional photo-editing, virtual staging, floor plan, and rendering service that can start with an app on your smartphone.

4. Image Recognition and Enhancement

Restb.ai: A tool specifically for the real estate market. It uses image recognition to categorize and tag photos of properties, making listings more searchable.

Styldod: Enhances property photos, online for free.

5. Lead Generation and Marketing Automation

Zurple: This platform uses AI to analyze the behavior of leads on your website. It then tailors automated responses based on the lead's activities.

Riley: An AI-driven text messaging system that qualifies leads round-the-clock.

6. Market Analysis and Trends

Sparktoro: Uses AI to analyze what a target audience reads, watches, shares, and discusses online. Helpful for real estate agents in targeting their marketing efforts.

7. CRM (Customer Relationship Management) Integration

KV Core: Leverage powerful smart AI automation to engage more contacts and close more deals.

Follow Up Boss: An AI-driven CRM that ensures leads are immediately contacted and nurtured accordingly.

8. Document Automation and Management

DocuSign: Uses AI to help real estate agents and their clients understand and act on agreement terms, speeding up the closing process.

SkySlope: An AI-enhanced platform for transaction management, ensuring compliance and keeping all documents in one place.

The real estate industry has seen a significant infusion of AI tools that cater to various needs, from lead generation to closing deals. Adopting these tools can streamline operations, enhance client interactions, and give real estate agents a competitive edge in a dynamic market. However, always ensure that any tool adopted aligns with your specific business needs and objectives.

Training and Support for Implementing AI Solutions

The real estate industry is no stranger to technological advancements. From online listings to virtual tours, technology has played a pivotal role in reshaping the sector. Now, with AI entering the fray, the landscape is shifting once again. However, integrating AI isn't as simple as installing new software; it requires proper training and continuous support. Here's a guide to ensuring a smooth transition.

1. Understand the Basics of AI

Before delving into specific tools, it's essential to grasp the fundamental concepts of AI. From the very basic to the more complex, the greater your understanding of the AI landscape, the greater your likelihood of success.

Online Courses: Platforms like Coursera, Udemy, and Khan Academy offer beginner courses on AI. While real estate agents don't need to be experts, a basic understanding can go a long way.

Workshops & Webinars: Look for industry-specific training sessions. These are often tailored to real estate professionals and can provide actionable insights.

2. Choose the Right AI Tool

Not every AI solution is right for every agent or agency. Understand your business needs first.

Consult with Vendors: Before purchasing, engage with the product vendors. They can provide demos, answer questions, and ensure their solution aligns with your needs.

Read Reviews: See what other real estate professionals are saying. Feedback from peers can offer invaluable insights.

3. Onboard with Expert Help

Once you've selected a tool, don't go it alone. Seek expert guidance during the initial phases.

Vendor Training: Most AI solution providers offer initial training sessions. Make the most of these to get your team up to speed while understanding that vendor training and consulting may be biased to specific products and technologies.

Hire a Consultant: If you're looking for more tailored assistance, consider hiring an AI consultant specializing in real estate. They can provide personalized training and implementation strategies.

Attend Seminars and Conferences: These events provide insights into the latest AI trends and offer practical workshops.

4. Ongoing Support

The AI journey doesn't end once the tool is up and running. Continuous learning and support are crucial.

> **Join User Groups**: Platforms often have user communities where members discuss challenges, share solutions, and provide support.
>
> **Stay Updated**: AI is ever-evolving. Regularly check in with your software provider or user community to stay updated on new features or changes.

5. Internal Knowledge Sharing

Promote a culture of shared learning within your organization.

> **Regular Training Sessions**: Hold monthly or quarterly training sessions where team members can share their experiences, challenges, and solutions.
>
> **Workshop Sessions**: Organize hands-on workshops where agents can practice using the tools in real-time scenarios.
>
> **Mentorship Programs**: Pair less tech-savvy agents with those more comfortable with AI. This peer-to-peer learning can be incredibly effective.

6. Emphasize the Importance of Regular Updates

Build expectations and plan for changes and improvements.

> **Software Updates:** These can bring in new features or changes to existing ones.
>
> **Training for New Features:** Whenever a significant update occurs, organize a training session focusing on the new features.

7. Feedback Loop

AI tools thrive on feedback. Ensure there's a mechanism to relay user experiences back to the vendors.

Scheduled Check-ins: Set up regular intervals, like every six months, to review the tool's performance and provide feedback.

Open Communication Channels: Encourage your team to report any issues or challenges they face. This feedback can be invaluable for both internal improvements and vendor communications.

Iterative Training: Use feedback and experience to design further training sessions, ensuring that real, on-ground challenges are addressed.

Integrating AI into real estate practices may initially appear challenging, yet with appropriate training and support, it transforms into a feasible and advantageous undertaking. It's important to recognize that the objective extends beyond merely embracing a new technology. The true aim is to leverage AI's capabilities to enhance service quality, make well-informed decisions, and maintain a competitive edge in the dynamic real estate industry. The key to effectively deploying AI lies in solid training and consistent support, which are critical for its successful implementation.

The Future of Real Estate with AI

Potential Advancements in AI for Real Estate: The real estate landscape is on the brink of a transformative shift, courtesy of Artificial Intelligence. While we have only scratched the surface of AI's potential in the field, the future beckons with innovations that could redefine property transactions, customer interactions, and overall business strategies. In this chapter, let's journey into the future and envision what AI has in store for the real estate sector.

1. Enhanced Predictive Analytics

While today's AI systems can predict housing market trends based on historical data, future advancements will allow for more granular, hyper-local predictions. Imagine AI tools that can predict the price fluctuations of a single property over the next decade, or which upcoming local infrastructure projects will impact property values.

2. Virtual Reality (VR) Property Tours Powered by AI

Beyond the basic 3D tours, AI-driven Virtual Reality (VR) will offer prospective buyers personalized tours. For instance, if a buyer has shown interest in spacious kitchens during past viewings, the AI-driven VR tour will highlight the kitchen features of a property first.

3. AI-Powered Building Design and Construction

In the construction domain, AI can assist architects in designing structures optimized for energy efficiency, space utilization, and even local weather conditions. By analyzing vast data sets, AI could suggest designs that are both aesthetically pleasing and functional.

4. Intelligent Property Management Systems

Future property management systems will be AI-driven, automating tasks like tenant communication, maintenance scheduling, and rent collection. These systems could even predict when a property is likely to need repairs.

5. Advanced Chatbots for Instant Customer Service

While chatbots are prevalent now, the next generation will be indistinguishable from human agents in conversation, providing instant, accurate, and personalized responses to complex queries around the clock.

6. Automatic Property Matching

Current property search engines rely on user-defined filters. The future AI-driven platforms will observe user behaviors, preferences, financial capabilities, and even intangible factors like emotional responses to create a precise property match, often finding listings the buyer hadn't even considered.

7. Enhanced Customer Relationship Management (CRM) Systems

The CRM systems of the future will be proactive rather than reactive. They will notify real estate agents when a client is likely ready for a property change based on life events, financial changes, or market conditions, ensuring real estate agents can offer services precisely when needed. The CRM of the future will work with AI to create a complete picture of a client's needs.

8. AI-Driven Legal and Documentation Assistance

One of the most challenging aspects of real estate is the paperwork. Future AI tools will streamline this process, auto-filling information, ensuring legal compliance, spotting discrepancies, and even predicting potential legal issues before they arise.

9. Emotion Recognition for Property Showings

While it may sound like science fiction, emotion recognition technology is advancing rapidly. For real estate agents, this could provide invaluable feedback. AI tools might analyze prospective buyers' facial expressions during a house showing, offering insights on which aspects of a property truly resonate with clients.

10. AI in Mortgage Processing

The mortgage approval process, known for its complexity, will be streamlined with AI. Advanced systems will assess an applicant's financial health, property value, and market conditions to make instant loan decisions.

11. Enhanced Security with AI Surveillance

AI-driven surveillance systems in properties will not just record activity but will predict and prevent unauthorized access by analyzing patterns, recognizing unfamiliar faces, and even detecting suspicious behaviors.

12. Sustainable Real Estate Practices

With climate change concerns on the rise, AI will assist real estate agents and builders in developing sustainable properties. It will analyze environmental factors to suggest the best renewable energy sources, insulation types, and more, aiming for net-zero energy consumption.

13. Integration of IoT with AI

Imagine a house that learns from you. With the Internet of

Things (IoT) devices powered by AI, homes in the future will adjust lighting based on the time of day, regulate temperatures based on occupancy, and even reorder groceries or household supplies when they run low.

14. Evolution in Property Marketing

AI will create hyper-personalized advertising campaigns. By analyzing potential buyers' online behaviors, preferences, and past interactions, it will curate property listings tailored just for them.

15. Real-time Language Translation and Global Transactions

For real estate agents dealing with international clients, AI-powered real-time translation tools will ensure smooth communication. Moreover, AI can simplify international property transactions by predicting forex trends, understanding global market shifts, and ensuring legal compliance across borders.

The real estate landscape is on the cusp of an AI-driven revolution, promising not only to make transactions smoother but also to make them more transparent, personalized, and insightful. As real estate agents, embracing these advancements will not only offer competitive advantages but will also redefine the very essence of property buying and selling. The future of real estate is not just about brick and mortar; it's about bytes, data, and intelligent algorithms.

Stay ahead, stay informed, and be ready to harness the immense potential of AI in real estate!

Preparing Your Business for an AI-Driven Future

As the real estate industry steadily pivots towards an AI-driven paradigm, it's imperative for real estate agents to stay ahead of the curve. Embracing AI doesn't mean merely adopting the latest tools; it's about cultivating a mindset ready for change. Here's a roadmap to prepare your real estate business for this imminent AI revolution:

1. Educate and Embrace

Stay Updated: The AI landscape is rapidly evolving. Attend seminars, webinars, and workshops focused on AI in real estate. Knowledge is your first tool.

Combat Fear with Facts: Understand that AI is not here to replace you, but to enhance your capabilities. Be open to its potential and benefits.

2. Invest in the Right Tools

Research and Choose: There's an ever-growing number of AI-driven tools for real estate. Invest in ones that align with your business needs – be it CRM systems, predictive analytics platforms, or virtual tour solutions.

Prioritize User-Friendliness: The best AI tool is the one you and your team can use easily. A steep learning curve can deter adoption.

3. Data Management

Prioritize Data Quality: AI thrives on quality data. Ensure that your property listings, client information, and other business data are accurate and up-to-date.

Understand Data Privacy: Familiarize yourself with data privacy laws and ensure that your AI tools comply with them. Be transparent with clients about how, when, and by who their data will be used.

4. Redefine Customer Interactions

Personalize with AI: Use AI tools to provide hyper-personalized services to clients, from property suggestions to communication strategies.

Balance Technology and Humanity: While AI can handle many tasks, the human touch remains irreplaceable in real estate. Use AI to handle data and analytics, but retain the personal touch in negotiations and client relationships.

5. Skill Upgradation

Training: Ensure you and your team are trained to use the AI tools you adopt. This not only boosts efficiency but also confidence in the technology.

Diversify Skills: With routine tasks automated, real estate agents should focus on enhancing skills that AI can't replicate – like negotiation tactics, relationship building, and local market expertise.

6. Adapt Marketing Strategies

Hyper-Targeted Marketing: Utilize AI algorithms to target potential clients with pinpoint accuracy, ensuring higher conversion rates.

Analyze and Adapt: Use AI-powered analytics to understand the effectiveness of your marketing campaigns. Adjust strategies in real-time based on this feedback.

7. Integrate Across Platforms

Ensure your AI tools can seamlessly integrate with other platforms you use, such as property listing sites, social media, and email marketing tools. This creates a unified and efficient ecosystem for your business operations.

8. Feedback and Iteration

Adopt a mindset of continuous improvement. Regularly gather feedback from your team and clients about their experience with your AI tools. This will guide you in making necessary tweaks and adaptations.

9. Plan Financially

While AI tools promise significant returns in terms of efficiency and client conversion, they also require investment. Plan your finances accordingly, and consider it an investment in your business's future.

10. Stay Connected with the AI Community

Join forums and communities that discuss AI in real estate. This keeps you in the loop about the latest advancements and best practices and offers networking opportunities.

The AI wave in real estate is more of an opportunity than a challenge. By preparing adequately, real estate agents can harness AI's power to offer unparalleled services, make informed decisions, and stand out in a competitive market. Remember, in the AI-driven future of real estate, it's not just about having advanced tools but about integrating them into your business fabric in a way that amplifies human potential.

Embracing the Future with AI in Real Estate

The evolution of Artificial Intelligence (AI) across various industries has been undeniable, and as we have explored throughout this book, its impact on the real estate sector is both profound and promising. In a world where information is abundant and the pace of technological advancement is rapid, the real estate industry cannot afford to remain static. AI is not just another tool—it represents a paradigm shift in how business is conducted, how decisions are made, and how value is delivered to clients.

From streamlining property searches to offering precise market analysis, the applications of AI in real estate are vast and varied. Virtual tours and augmented reality have transformed property viewings into immersive experiences, while AI-driven Customer Relationship Management tools (CRMs) are revolutionizing client relationship management, enabling real estate agents to offer personalized and efficient service.

But with great power comes great responsibility. As we leverage AI to its fullest potential, it is essential to remain aware of the ethical considerations. Data privacy, algorithmic biases, and the balance between the human touch and automation are all pertinent topics that real estate agents must address. It's not just about harnessing AI's capabilities but using them wisely and ethically.

As real estate agents, embracing change and innovation Artificial Intelligence is not optional—it's imperative. The journey into the world of AI might seem daunting, but it's an exciting one, filled with endless possibilities. The future is not something we merely predict; it's something we shape. By harnessing the power of AI and committing to lifelong learning, real estate agents can be at the forefront of this transformation, leading the charge into a new and promising era of real estate.

In conclusion, AI in real estate is not just a trend—it's the next frontier. As we stand on the brink of this AI revolution, let us step forward with curiosity, diligence, and a commitment to excellence. The future is bright, and it's time to shine.

Thank you for joining us on this enlightening journey. As the realm of AI continues to expand, may your passion for real estate and technology guide you towards success and fulfillment.

Appendix A

AI Model Development Prompt for Spotting the Next HotSpot with AI.

Objective:
Develop an AI model designed to assist investors and real estate agents in identifying emerging property hotspots by analyzing and predicting neighborhood growth patterns, ultimately enabling data-driven investment decisions.

Input Data:

Neighborhood Profiles: [Area Name, Historical Property Values, Past Growth Patterns, etc.]

Socio-Economic Indicators: [Population Growth, Income Levels, Employment Rates, etc.]

Infrastructure Developments: [Upcoming and Completed Projects such as Transportation, Shopping Malls, Schools, etc.]

Emerging Businesses: [Startups, New Shops/Restaurants, Business Migrations, etc.]

Policy Changes: [New Government Policies, Tax Implications, Zoning Laws, etc.]

Scenario: Two neighborhoods are under consideration for property investment - A (historically popular) and B (experiencing modest growth). Your AI model will evaluate and predict the future growth and popularity of these areas by meticulously analyzing the provided input data, thereby assisting in identifying which neighborhood might yield better future returns on investment.

Model Specifications:

Algorithm Type: [Decision Trees, Random Forest, Gradient Boosting, etc.]

***Data Preprocessing:** [Handling Missing Values, Encoding Categorical Variables, etc.]

Feature Engineering: [Combining Variables, Extracting Insights, etc.]

Model Training: [Defining Training & Validation Sets, Hyperparameter Tuning, etc.]

Model Evaluation: [Accuracy, Precision, Recall, F1 Score, etc.]

Expected Output:

Growth Predictions: [Predicted Growth Rate, Popularity, Property Values, etc.]

Investment Recommendations: [Where and When to Invest for Optimal Returns]

Risk Analysis: [Potential Risks and Mitigations in Investing in Areas A and B]

Visualizations: [Charts and Graphs Visualizing Growth Predictions and Comparisons]

Explanation: [Detailed Insights into Why Area B might Become a Hotspot]

User Interface:

Inputs: [Options for Users to Input/Upload Data about Various Neighborhoods]

Results Display: [Visual and Textual Display of Predictions and Recommendations]

Feedback Mechanism: [Option for Users to Provide Feedback on Predictions]

Security and Compliance: Ensure all data utilized and generated by the model adheres to local and international data protection regulations. Implement robust cybersecurity protocols to safeguard user data.

Additional Notes:

Usability: Ensure that the model and its interface are user-friendly and accessible to individuals with varying degrees of technical expertise.

Continuous Learning: Incorporate mechanisms to continually learn and adapt from new data and user feedback, ensuring the model remains accurate and relevant.

Scenario Implementation: Utilizing the AI model, investors/agents will analyze the current and predicted future states of neighborhoods A and B. While A has consistently demonstrated popularity, the model will scrutinize various indicators and historical data from B to determine if it's poised to become a lucrative investment opportunity in the near future. By providing detailed insights, visualizations, and explanations, the AI assists users in identifying potential under-the-radar investment opportunities, thereby enabling them to strategize their investments meticulously and maximize future returns.

Appendix B

AI Model Development Prompt for Predicting the Best Time to Sell

Objective: Develop an AI model capable of predicting the price trends of properties in bustling urban areas by analyzing various micro and macro factors, thereby assisting property owners, investors, and real estate professionals in making informed decisions regarding holding or selling properties.

Input Data:

Property Profiles: [Location, Size, Type, Historical Price, etc.]

Future Development Projects: [Public and Private Projects, Infrastructure Development, etc.]

Political Stability: [Election Outcomes, Political Trends, etc.]

Employment Rates: [Job Availability, Company Headquarters, etc.]

Policy Changes: [Upcoming Legislation, Zoning Laws, Tax Regulations, etc.]

Scenario: A property is situated in a highly active urban environment, conventionally associated with continuous price elevation. The AI model will delve into a comprehensive analysis, considering numerous factors that could influence future price trajectories, potentially identifying opportunities or predicting risks such as plateaus or declines in property prices, providing crucial insights for decision-making processes regarding property management and investment.

Model Specifications:

Algorithm Type: [Linear Regression, Decision Trees, Neural Networks, etc.]

Data Preprocessing: [Handling Missing Values, Normalization, etc.]

Feature Engineering: [Creating Interaction Terms, Variable Transformations, etc.]

Model Training: [Data Splitting, Cross-Validation, Hyperparameter Tuning, etc.]

Model Evaluation: [Mean Absolute Error, R-squared, Root Mean Squared Error, etc.]

Expected Output:

Price Predictions: [Future Price Estimates, Price Trend, etc.]

Recommendations: [Whether to Sell Now, Wait, or Other Strategic Movements]

Risk Analysis: [Potential Risks Involved in Each Decision]

Visualizations: [Graphs and Charts Displaying Predicted Trends and Factors Influencing Prices]

Explanation: [Insights into What Factors Led to the Predictions and Recommendations]

User Interface:

Inputs: [Options for Users to Enter or Upload Property and Macro-factor Data]

Results Display: [Visually and Textually Convey Predictions, Recommendations, and Insights]

Feedback Mechanism: [Allow Users to Provide Feedback on Predictions and Recommendations]

Security and Compliance:
Ensure data privacy and compliance with applicable data protection legislation.
Implement cybersecurity best practices to protect user data and model outputs.

Additional Notes:
Usability: The model interface should be intuitive, facilitating ease of use for individuals with various levels of technical expertise.
Continuous Learning: Integrate mechanisms for the model to adapt and enhance its predictive capabilities continuously utilizing new data and feedback.

Scenario Implementation: For a property owner in a vibrant urban area, the AI model will dissect various influencing factors to predict the future pricing trend of the property. Despite conventional beliefs about perpetual price growth in active urban zones, the model could identify potential future stabilizations or declines in property values based on analyzed factors, thereby aiding owners and investors in deciding whether to capitalize on current values or await potential future appreciations.

Appendix C

Navigating through the Busy City Markets with AI-Enhanced CMA

Urban real estate markets are notorious for their ever-changing nature, with property values fluctuating due to myriad factors, including new development projects, economic changes, and shifts in policy. Traditional Comparative Market Analysis (CMA), which evaluates property values based on recent sales of similar properties, might not fully capture these dynamic factors. This is where Artificial Intelligence (AI) steps in, offering a solution that is not only deeply analytical but also predictive, considering the present scenario and future potential changes, ensuring properties are priced aptly, both for the current market conditions and anticipated future shifts.

Scenario 1: Accommodating Developmental Projects in Property Valuation

Imagine a residential property nestled in the heart of a thriving city. This property is affected not only by its intrinsic value but also by a multitude of external factors. For example, a new shopping mall or a park development project in the vicinity is likely to impact its value.

In a traditional CMA, you might assess the property's value based on the sale prices of similar properties recently sold in the area, possibly overlooking the potential impact of the new development. However, by integrating AI, you can leverage predictive analysis to factor in the likely impact of the new development on the property's future value. The AI would analyze data from similar past scenarios, predict the potential impact of the development, and incorporate this into a more dynamic, future-oriented pricing strategy for the property.

Scenario 2: Adapting to Economic and Policy Shifts

Consider another property in the city, where an economic downturn or a new government policy (such as a change in property tax or zoning laws) is poised to affect property values in the area. Traditional CMA might not fully account for these changes, especially if their impacts have not yet been realized in the sale prices of comparable properties.

AI, on the other hand, can be trained to recognize patterns and predict outcomes based on similar historical events, providing a pricing strategy that's proactive rather than reactive. For instance, if an economic downturn is anticipated, AI could analyze data from previous downturns, predicting potential impacts on property values and allowing for a pricing strategy that anticipates this future change, keeping the pricing competitive and in line with future market expectations.

In both scenarios, AI complements the traditional CMA by providing a pricing strategy that's not just based on the present conditions, but also anticipates and adapts to potential future changes, ensuring that pricing is robust, competitive, and strategically sound in the face of the dynamic urban market's fluctuations.

By embracing AI's predictive capabilities, real estate agents can navigate the bustling and ever-changing urban markets with a pricing strategy that is as dynamic and flexible as the markets themselves. This allows for pricing that is competitive today while also being prepared for tomorrow, offering a significant edge in the fast-paced world of urban real estate.

Appendix D

Analyzing a Suburban Property with Historical Significance Using AI-enhanced CMA

Suburban properties with historical significance pose a unique challenge in valuation. Their value isn't only determined by size, condition, and location but is often substantially influenced by their historical aspect, which might be nuanced and multifaceted, affecting value in ways that traditional Comparative Market Analysis (CMA) might not fully encapsulate. Let's delve into a scenario where AI could be significantly impactful in deriving an apt valuation for such a property.

Scenario: Valuing a Suburban Property with Rich Historical Backdrop

Envision a quaint property located in the suburbs, but this isn't just any property – it's a home that's embedded with rich historical significance, perhaps being the birthplace of a notable person or a site where a significant event took place. Such a property isn't merely a structure; it's a piece of history, potentially attracting tourists, historians, and those who have a keen interest in cultural heritage. Conventional CMAs might struggle to quantify the additional value that this historical significance brings to the property.

By introducing AI into the evaluation process, real estate agents can enhance their CMA to capture the nuanced value that historical significance can bring to a property. AI can examine extensive datasets, diving deep into the impact that historical value has had on similar properties in different regions. It will look into variables like tourist attraction levels, the cultural importance of the site, and any policies related to the conservation of historical sites, providing a much richer, contextual valuation.

The AI would analyze how similar historically significant properties have been valued and sold in the past, considering not just the raw data of sale prices, but also contextual data related to the aforementioned variables. By analyzing this data, AI could predict how these factors might influence the value of the property in question and provide a valuation that isn't merely a reflection of its physical attributes and recent sale prices of comparable properties, but a valuation that's informed by the additional, nuanced value that its historical significance brings to the table.

In this scenario, AI doesn't replace the CMA; it enhances it. The advanced analytical capabilities of AI ensure that the valuation is not only contextually accurate but also nuanced, reflecting the property in a manner that is intrinsically tied to its unique characteristics, thereby offering a market-relevant valuation that is truly representative of the property in all its historical grandeur.

In the constantly evolving world of real estate, integrating AI with traditional practices like CMA ensures that valuations are not only accurate but also intimately tied to the unique stories and contexts of each property, particularly when dealing with properties that boast of a rich historical backdrop, ensuring a well-rounded, and robust approach to property valuation.

By enabling a deeper, more nuanced approach to property valuation, AI empowers real estate agents to navigate the complex waters of valuing historically significant properties, ensuring their unique characteristics are fully acknowledged and accurately reflected in their market valuation.

Appendix E

Here's a list of AI-driven CRMs that have gained traction in the real estate industry:

Zoho CRM: Incorporates Zia, Zoho's AI-powered assistant, which analyzes data to provide sales predictions, lead scoring, and even sentiment analysis.

kvCORE: Designed specifically for real estate, it offers behavioral automation, lead gen tools, and business analytics driven by AI.

Chime: Their AI-driven system helps agents target leads more efficiently, track buyer activities, and offers chatbots for instant responses.

Real Geeks: Offers an AI-driven lead nurturing system through automated emails and texts based on lead behavior.

LionDesk: Features an AI-powered transaction management system and video marketing tools.

Propertybase: Integrates AI to drive business processes, engagement, and grow opportunities for real estate brokers and teams.

Follow Up Boss: Automates lead distribution and follow-ups based on AI-driven insights.

REthink CRM: Built on Salesforce platform, it leverages AI for data-driven insights to help agents and brokers optimize their strategies.

BoomTown: Features a predictive CRM that uses AI to help agents prioritize and automate tasks.

Remember, the best CRM for your real estate business will depend on your specific needs, size, and budget. It's always a good idea to request demos, read reviews, and compare features before making a final decision.

Appendix F

Here are several companies that provide AI chatbots specifically designed for the real estate industry:

Structurely (Holmes): This AI assistant specializes in qualifying leads, ensuring real estate agents communicate with the most promising prospects.

Roof AI: A platform that connects potential buyers with listings by having conversations through chatbots.

Zenplace: Uses AI-powered chatbots to provide property management solutions, assisting both landlords and tenants.

Riley: Provides 24/7 lead qualification services, making sure real estate agents never miss out on potential opportunities.

Chatfuel: While not exclusively for real estate, Chatfuel is a platform where agents can create AI chatbots for Facebook Messenger to engage with potential leads.

HelloAlex: An automated income assistant for real estate agents that engages leads within 2 minutes of their inquiry.

Offrs: Uses AI to predict property listings before they even come to market, and comes with a chatbot feature for lead engagement.

KwizCom: Offers a chatbot that assists with property search on real estate websites.

ZoConvert: Another tool not exclusively for real estate, but allows agents to build chatbots for Facebook Messenger to help with lead generation.

When considering any AI chatbot solution, it's crucial for real estate professionals to research each option thoroughly, understand the costs involved, and ensure the chosen solution aligns with their business goals and target audience.

Appendix G

Sample Privacy Policy with AI Language

1. Introduction
Welcome to [Your Company Name]. We value your privacy and are committed to protecting your personal data. This privacy policy outlines how we collect, use, store, and protect your personal data, especially in light of our AI-driven solutions.

2. What data we collect
When you use our services, we may collect:
Basic details: Name, email, address, phone number.
Behavioral data: Your preferences, feedback, and interactions with our website or app.
Technical data: IP address, device type, operating system, browser type.

3. How we use your data
We utilize your data to:
Improve our services.
Personalize your experience.
Process transactions or requests.
Train our AI models to better serve you. This helps the AI understand general user preferences and behaviors but does not specifically identify or target individual users.

4. How we use AI
Our AI systems analyze the data to:
Predict and suggest solutions or products you might be interested in.
Automate repetitive tasks and improve efficiency.
Enhance user experience based on patterns and trends.

5. Data sharing and storage
We do not sell your personal data. However, we might share your

data with trusted third-party services that help us run our platform. All data, when used for AI training, is anonymized, ensuring no specific individual can be identified from the AI's knowledge.

6. Your rights
You have the right to:
Access the personal data we hold about you.
Request corrections of any inaccurate data.
Delete your personal data from our systems.
Opt-out of certain data processing activities.

7. Data security
We have implemented security measures to ensure your data is safe. This includes encryption, secured servers, and regular security assessments.

8. Cookies
We use cookies to enhance your experience, gather general visitor information, and track visits to our website. You can choose to disable cookies through your browser settings.

9. Changes to this policy
We may update this policy from time to time. We recommend reviewing it regularly to stay informed.

10. Contact us
If you have any questions regarding this privacy policy or our use of AI in handling your data, please contact us at [Your Email Address].

Note: This is a generic sample privacy policy and might not cover all legal requirements. Always consult with a legal expert when crafting a privacy policy for your business.

Made in United States
Troutdale, OR
05/29/2024